Fundamental Physics of Sound

Fundamental Physics of Sound

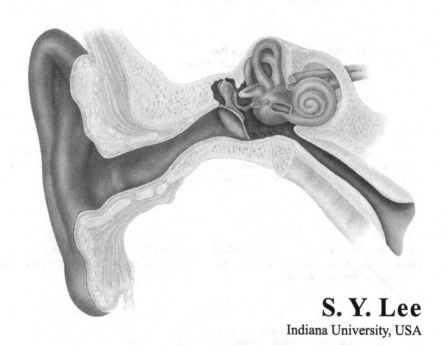

S. Y. Lee
Indiana University, USA

World Scientific

NEW JERSEY · LONDON · SINGAPORE · BEIJING · SHANGHAI · HONG KONG · TAIPEI · CHENNAI · TOKYO

Published by

World Scientific Publishing Co. Pte. Ltd.

5 Toh Tuck Link, Singapore 596224

USA office: 27 Warren Street, Suite 401-402, Hackensack, NJ 07601

UK office: 57 Shelton Street, Covent Garden, London WC2H 9HE

Library of Congress Cataloging-in-Publication Data

Names: Lee, S. Y. (Shyh-Yuan), author.

Title: Fundamental physics of sound / S.Y. Lee, Indiana University, USA.

Description: New Jersey : World Scientific Publishing Co., Inc., [2020] |
    Includes bibliographical references and index.

Identifiers: LCCN 2020036678 (print) | LCCN 2020036679 (ebook) |
    ISBN 9789811222597 (hardback) | ISBN 9789811222603 (ebook)

Subjects: LCSH: Sound.

Classification: LCC QC220 .L45 2020 (print) | LCC QC220 (ebook) | DDC 534--dc23

LC record available at https://lccn.loc.gov/2020036678

LC ebook record available at https://lccn.loc.gov/2020036679

British Library Cataloguing-in-Publication Data

A catalogue record for this book is available from the British Library.

For any available supplementary material, please visit
https://www.worldscientific.com/worldscibooks/10.1142/11893#t=suppl

Printed in Singapore

Preface

The idea of this book came from my teaching experiences in two related courses, intermediate acoustic lab and the basic physics of sound. In 2009, I began teaching the "acoustic laboratory" course at the Physics department, Indiana University, Bloomington. In 2013, I taught the "Basic Physics of Sound," an elective course for general education designed for students majoring in music, speech and hearing, and audio technology. Attempting to facilitate students' learning, I prepared extensive course notes for them. At the same time, I worked hard to understand what learning difficulties students might have encountered. I realized that it was mostly psychological. They were generally afraid of math and physics (see Ref. [1] for basic math preparation). In fact, math is nothing but logical procedure in getting physics information. This is a book derived from these lecture notes.

In teaching the basic physics of sound, I found that computer demonstrations are an important part of the learning. Fortunately, there are free audio-synthesis programs available online (see Refs. [2–3]). Furthermore, there are a number of existing good textbooks and acoustic handbooks on the "science of sound" (see Refs. [4–21]). However, most textbooks contain too much information that it is difficult to cover essential topics in a 16-week semester. I make special efforts to develop a set of lecture notes for the 16-week semester course. My aim is to produce a textbook, which is about 300 pages long, divided into 4 parts that fit into the semester structure, containing enough latest information on the subject, and that is economically affordable to students.

The book is formulated in 4 parts. The first part deals with basic Newton's second law of motion, simple harmonic oscillation, and wave properties. Newton's second law is $F = ma$, i.e. the net force is equal to the product of mass and acceleration. Newton formulated his second law as "the net force is equal to the rate of change of momentum." I use this formulation to derive the speed of waves in a medium. This is a much more satisfactory derivation for the wave speed than what we have seen in the literature.

The second part of the textbook focuses on the three attributes of sound: loudness, pitch and timbre. This part includes the neurobiology of our hearing system, the interplay between the place and temporal theories of pitch, the psychoacoustics in our perception, and musical scales.

The third part of the book discusses the basic physics of some musical instruments and human voice (see Refs. [22–31]). From the point of view of physics, musical instruments and human speech are similar. They are composed of a sound source and a resonator. Human ingenuity has produced various aesthetic-looking and ear-pleasing instruments for musicians to perform. Magical human evolution has also shaped our vocal folds and vocal tract so that we can dynamically change loudness, pitch, and timbre in an instant, in a manner that no other musical instrument can emulate.

The fourth part of the book includes electricity and magnetism, room acoustics, digital technology in acoustics, effects of noise on human hearing, and noise regulations for hearing protection. These topics are relevant to the production, transmission, and storage of sound, and to human ear protection. Our ears are an extremely sensitive instrument. Without proper protection, loud noise including loud music can damage our ears (see Refs. [32–40]). Government regulation and education serve as a first line of protection in working environments.

Technology advancement is progressing very fast. Synthesis of voice has already been available in information technology. The artificial intelligence (AI) systems can understand human language, process required information to communicate with people, and provide services to our need. The AI technology may someday compose music, perform music, and produce a synthetic "human" voice of some particular artist that we admire.

During the past few years, I have learned much from my experience in teaching this subject to about 750 students. These students taught me how to give a better explanation of various topics in this book. It is algebra based. The comprehension of the topics does not need the skills of calculus. Students do need to know some simple laws of mechanics, e.g. Newton's 2nd law of motion, Hooke's law on motion of objects away from equilibrium, and the Bernoulli principle. Beyond the physics, they also need to have a calculator with logarithmic function and sine-cosine functions.

There are homework problems at the end of each chapter. References to the original articles are included as in-text footnotes. I hope this arrangement will cause less distraction for students when they read the textbook. Answers to all homework problems are included in Appendix F. This may help some beginner students on this subject.

I may have achieved my objective of writing a small book. However, like many authors, I often wonder whether this book is useful to students. Are the contents presented in this book readable, comprehensible, understandable and enjoyable to students? Until this book is used in classrooms, I will not get answers to these questions. I will be delighted if this book provides good services to all eager students.

S.Y. Lee
Bloomington, Indiana, USA
July 2020

Acknowledgments

My interest in "physics of sound" arises from my teaching of "basic physics of sound" and the "intermediate acoustic laboratory" courses at the Physics department, Indiana University Bloomington, Indiana, USA. After the completion of the draft version of this book. I have benefited from a few colleagues, Ping Chou and Linh Nguyen, who help to read the manuscript and provide valuable suggestions. In particular, I am truly indebted to Linh Nguyen. Her help in editing the text is indispensable for the completion of the textbook. She helps to provide a proper structure in the explanation of physics concepts. I would like to give heartfelt thanks to the World Scientific editing team for their help in transforming the book into this final form and in catching equation misnumbering. The responsibility for all misunderstanding and errors lies with me. Please address all comments and corrections to me and your comments and corrections will be highly appreciated.

Contents

Abbreviation of terms in science of sounds

AC: alternating current

ADC: analog to digital conversion

AHL: age related hearing loss or presbycusis

AI: artificial intelligence

AJP: American Journal of Physics

AM: amplitude modulation

AN: auditory nerve

ANSI: American National Standards Institute

AOP: Acoustic Overload Point

AP: absolute pitch

ASA: Acoustical Society of America

ASIC: application-specific integrated circuit

ASTM: American Society for Testing and Materials

ASCII: American Standard Code for Information Interchange

ATM: atmospheric pressure 1.013×10^5 Pa

ATS: asymptotic threshold shift

BIL: band intensity level

CB(W): critical bandwidth

CNS: central nerve system

Codec: coding or decoding data stream

CD: compact disc

CF: characteristic frequency

CPU: Central Processing Unit

DAC: digital to analog converter

DC: direct current

dB: decibel

dBA: A-weighted decibel scale

dBB: B-weighted decibel scale

dBC: C-weighted decibel scale

DBFS: decibels relative to full scale

dBm: dB value re 1 mW

dBV: dB value re 1 V/Pa

DR: dynamic range

DVD: digital video disk

EGG: electroglottalgraph

EJPH: European Journal of Public Health

EM wave: electromagnetic wave

EM: electricity and magnetism

EMF: electromotive force (electric potential)

EMG: electromyography

EPNL: effective perceived noise level

ERB: equivalent rectangular band

FFT: Fast Fourier Transform

FM: frequency modulation

GBwP: gain-bandwidth-product

HL: hearing loss

HVAC: heating, ventilation, air-conditioning

IEC: International Electrotechnical Commission

IINSE: International Institute of Noise Control Engineering

ILD: Inter-aural Level Difference

IPA: International Phonetic Alphabet

ISO: International Organization for Standardization

ITD: Inter-aural Time Difference

JASA: Journal of the Acoustical Society of America

JAMA: Journal of American Medical Association

Jnd: just noticeable difference

KTH: Royal Institute of Technology; Stockholm, Sweden

LOG (log): logarithmic function based 10

LPCM: linear pulse code modulation

LSB: least significant bit

MEMS: micro-electro-mechanical system

MFP: mean free path

MKS: meter-kilogram-second in SI unit

MPEG: Moving Picture Experts Group

MRI: magnetic resonance imaging

MSB: most significant bit

NCA: Noise Control Act

NIH: National Institute of Health

NIHL: noise-induced hearing loss

NIDCD: National Institute on Deafness and other Communication Disorders

NIOSH: National Institute for Occupational Safety and Health

NIPTS: noise induced permanent threshold shift

NIST: National Institute of Standards and Technology

NRC: Nuclear Regulatory Commission

OAE: otoacoustic emission

Op-amp: operational amplifier

OSHA Occupational Safety and Health Administration

PCM: pulse code modulation

PEL: permissible exposure limit

PNL: perceived noise level

PNAS: Proceedings of the National Academy of Science USA

PTS: permanent threshold shift

Q-factor: quality-factor

QCA: Quiet Communities Act

REL: Recommended Exposure Limit

RLC: resistor, inductor, capacitor circuit

RMS (rms): root mean square, $\sqrt{\langle p^2 \rangle}$

SHM: simple harmonic motion

SHO: simple harmonic oscillator

SI unit: system international unit

SII: Speech Intelligibility Index

SIL: sound intensity level

SNR: signal to noise ratio

SPL: sound pressure level

SRI: sound reduction index

STC: sound transmission class

STI: Speech Transmission Index

STP: standard temperature and pressure

THD: total harmonic distortion

TTS: temporally threshold shift

TL: transmission loss

TWA: time-weighted averaged

UTF: Unicode Transformation Format

WHO: World Health Organization

Chapter 1

Classical Mechanics and Simple Harmonic Motion

Acoustics is an interdisciplinary science that studies properties of sound waves. Waves are disturbances of "physical quantities" that propagate in space and time. The physical quantities include "displacement and velocity" of molecules, "pressure," "electric and magnetic fields," etc. For example, the electromagnetic (EM) waves are oscillatory "electric and magnetic fields" traveling in space and time. A sound wave is a "pressure" disturbance that propagates in media, including gases, liquids and solids. The pressure disturbance depends on properties of the medium such as density, elasticity, geometry and equations of states of the physical systems. Newton's law of mechanics governs the mechanism of the disturbance and its propagation.

Waves carry energy and momentum. We characterize the sound wave by its power, intensity and frequency content, and we perceive sound wave through our hearing system in loudness, pitch and timbre. We communicate with each other through voice, language and music. This book covers only basic physics of sound, divided into four parts:

Part I (Chapters 1 and 2) includes classical mechanics dealing with kinematics and dynamics, Newton's law of motion, simple harmonic motion, standing waves or resonances, properties of waves, energy density and intensity of waves.

Part II (Chapters 3 and 4) addresses human hearing system; psychoacoustic perception on sound; and three attributes of sound: pitch, timbre, loudness, synthesis and musical scales.

Part III (Chapters 5 and 6) discusses sound production including musical instruments, and human voice production.

1

Part IV (Chapters 7 to 10) explores the basic electricity and magnetism in sound production, electric circuits, transducers such as microphones and loudspeakers, digital technology in sound storage, recording, room acoustics, effects of sound on human hearing system, noise control and noise regulations.

I.   What Is Sound?

The world is full of waves. Examples of waves are electromagnetic waves, sound waves, water waves, earthquake waves, waves on string, waves on a drum surface in music performance, etc. All waves transport energy and momentum from one point in space-time to another. The energy and momentum in waves come from the oscillation of "physical quantities." A sound wave is the oscillation of molecules as it propagates forward in media. A wave on a string is the up-down motion of the string medium. Waves transport wave energy and wave momentum without transporting materials in the medium.

The motion of many objects is periodic. Periodic motion will produce a periodic wave. A simple periodic wave is the "sinusoidal wave" with the period T and frequency $f = 1/T$. The frequency is the number of oscillations per second. As the sinusoidal periodic wave propagates in space, the distance for one complete oscillation in the medium is the wavelength λ. Thus, wave speed is $v = f\lambda$. All waves can be decomposed into sinusoidal waves at various frequencies (see Fourier theorem in Appendix B).

We generally classify waves into longitudinal and transverse waves. When the motion of the wave medium is perpendicular to the direction of wave propagation, it is a transverse wave. For example, the up-down motion of a piano string is perpendicular to the direction of wave propagation along the string's length. The EM wave is also a transverse wave because the oscillatory electric and magnetic fields are perpendicular to the wave propagation direction.

When the motion of the medium is in the same direction as the wave propagation, it is a longitudinal wave. For example, air molecules move forward and backward as a sound wave travels forward. As molecules in a medium move forward and backward, they produce "pressure" oscillations. *A sound wave is a pressure wave in media.* It causes the compression and rarefaction of molecules in the transmission medium. The motion of molecules obeys the laws of classical mechanics. Familiar sound waves are human voices, bird chirps, drumhead and piano sounds, etc.

All sound waves are produced by vibrating bodies. Sound waves travel in solid, liquid and gas (air), but not in vacuum. Molecules in the transmitting medium move forward and backward, and transport the acoustic energy-momentum forward, i.e. a sound wave is the propagation of a longitudinal pressure disturbance, or longitudinal pressure wave, through a medium. The speed of a sound wave in air is 343 m/s at 20°C on Earth.

We perceive, synthesize and interpret the sound wave in our hearing system, which includes our ear, auditory neurons, and the auditory cortex in the brain. Our perception of sound is termed "psycho-acoustic." We measure and characterize the sound wave using instruments. Modern computers are powerful tools for synthesizing and analyzing properties of sound waves. To understand the basic physics of sound, we begin with the "mechanics."

II. Classical Mechanics

Mechanics is the study of motion of physical objects. *Classical* mechanics is relevant to physics of sound because the velocity of particle motion is much slower than the speed of light, i.e. $v \ll c \equiv 299792458$ m/s $\approx 3 \times 10^8$ m/s. In contrast, *relativistic* mechanics deals with the motion of particles at speed approaching the speed of light. Since the sound wave involves the motion of a large number of molecules, we do not need *quantum* mechanics that deals with the motion one or few subatomic particles in atoms and nuclei.

Topics relevant to classical mechanics are (A) the units of physical quantities; scientific notation and significant figures; (B) kinematics involving displacement (loosely called distance); velocity; and acceleration; (C) dynamics of motion governed by Newton's laws of motion; and (D) work, energy, power and intensity.

A.   Units, Scientific Notation and Significant Figures

Units

Measurements of any physics quantity require a unit. For example, we measure the weight of grocery in pounds, mass in kilograms, distance in meters or feet, and time in seconds. The SI (*Système international*, or International System) unit system uses MKS [meter-kilogram-second], where "M" stands for [meter] or [m]; "K" for [kilogram] or [kg]; and "S" for [second] or [s].

In the USA, people also measure length in feet (singular: foot), yards, and miles; mass in "weight" with ounces, pounds, or tons; and time in seconds, minutes, hours and years. Conversion factors for these units are 1 inch = 2.54 cm, 1 m = 3.28 feet [ft], 1 foot = 0.3048 m, 1 mile = 1,609 m, 1 pound = the weight of a 0.454 kg mass on Earth, etc. Other societies use different measurement systems. In acoustics, we use SI units.

Scientific Notation, Significant Figures; Prefix of Large and Small Number

Science needs to deal with very large or very small numbers. It is cumbersome to write a number with many zeros, and thus we try to express the number in scientific notation. For example, the human body containing about 100 trillion (100,000,000,000,000) cells is expressed as 1×10^{14} cells; the typical human cell size is 0.00001–0.0001 m, expressed as 10^{-5}–10^{-4} m. The scientific notation of 0.000939 is 9.39×10^{-4} or 9.39E−4 when used as input/output for calculators and computers.

All scientific instruments and measurements have their precision limit and uncertainty. For example, when we measure a distance of 2.0 m, the actual distance may be 1.95 m up to 2.04 m. The number of significant figures of 2.0 is 2. The distance 2.00 m may represent 1.995 m up to 2.004 m, i.e. it has 3 significant figures. However, 2,000 m may represent 2×10^3 m or 2.0×10^3 m or 2.00×10^3 m for possibly 1, 2 or 3 significant figures. The number of significant figures represents the maximum trusted scientific number in a measurement. The International Astronomical Union set the speed of light (EM wave) in

"vacuum" at 299,792,458 m/s, exactly. The time in second is the oscillations of Cs-133 atom atomic clock 9,192,631,770 periods. The time has
significant figure of 10, and the meter is the distance that light travels
in 1/299792458 s.

The number of significant figures is important in arithmetic operations (addition, subtraction, multiplication or division) in sciences.
Some examples are

a) $7.3 + 2.24 = 9.5$ 2 significant figures (sometimes written as
sig figs or sf)

b) $7.3 + 200.24 = 207.5$ 4 significant figures

c) $7.3 \times 2.24 = 16$ 2 significant figures

d) $7.3/2.24 = 3.3$ 2 significant figures

The expression of big or small numbers in scientific notation is sometimes cumbersome. We also use prefix to represent these numbers. For
example, the typical human cell is about 10–100 μm, where μ is the
prefix of 10^{-6}. A typical computer Central Processing Unit (CPU) uses
14 nm technology in 2017 and 5–7 nm in 2019. The size of an atom is
typically about 0.25 nm. The Table below lists some typical prefixes.

Small numbers	c [centi-]	m [milli-]	μ [micro-]	n = [nano-]	p = [pico-]	f = [femto-]
	10^{-2}	10^{-3}	10^{-6}	10^{-9}	10^{-12}	10^{-15}
Big numbers	k = [kilo]	M = [mega]	G = [giga]	T = [tera-]	P = [peta]	E = [exa-]
	10^{3}	10^{6}	10^{9}	10^{12}	10^{15}	10^{18}

B. Kinematics

Kinematics in mechanics is the "state of motion." At any instant
of time, an object has a position in space. If the object moves, it
"displaces" from its original position to another position in space. The
change in position is the *displacement*. The rate of change in displacement is *velocity* (loosely called speed); and the rate of change in velocity
is *acceleration*.

Figure 1.1: (a) The map of a city block represents a coordinate system. If one chooses East to be positive, then West will be negative. (b) Schematic drawing of the position vs time showing the movement of a person from the "origin" in a total time interval of 14 s. The person strolls to the left (assuming the positive direction is to the right) to -4 m in 2 s; stays there for 2 s; walks to the right to $+4$ m from the starting position; stays there for 2 s; and walks back to the starting position in 4 s. The graph represents the state of the person's motion during the 14-second time interval.

1) Displacement

The displacement is the change in position in space. Figure 1.1(a) shows an example of a street map for a street in Bloomington, Indiana. It represents a 2-dimensional coordinate system. If you walk from the corner of S Jordan Ave and E 3rd Street 1 mile eastward, your displacement is 1 mile or 1.609 km eastward. If you walk 1 mile westward, then your displacement is 1 mile westward. If you define eastward as positive, then your displacement eastward is $+1.609$ km, and your displacement westward is -1.609 km. There is a "direction" associated with displacement. If we walk northward along the S Jordan Ave, the displacement is northward. If we walk 1 mile eastward and 1 mile northward, the displacement is 1.414 mile northeast (Pythagorean Theorem), and yet our total travel distance is 2 miles.

2) Velocity

The velocity is the rate of position change. The displacement is the change of position and thus the average velocity is displacement divided by time during this displacement:

$$v_{ave} = \langle v \rangle = \frac{\Delta x}{\Delta t} = \frac{x(t_2) - x(t_1)}{t_2 - t_1}, \qquad (1.1)$$

where Δx is the displacement and Δt is the time interval for this displacement. If the measuring time interval Δt is small, the average velocity becomes the *instantaneous velocity*. The unit of velocity, defined the change of position per unit time is [Length/Time]; or [m/s] in SI unit. In USA, we sometimes use miles/hour [mi/h] for the velocity. We loosely use speed (the absolute value of velocity) to represent velocity, where the average speed is $\langle v \rangle =$ (total distance)/time. The speed is always a positive number. A wave travels from a source outward in "all directions," thus; we call "wave speed," instead of wave-velocity. The speed of sound wave is 343 m/s in air at 20°C.

★ Example 1.1: If it takes 15 minutes to run 1.609 km westward, what is the average velocity if we choose + direction to be eastward?

Answer: The average velocity is $-\frac{1609\,\text{m}}{15 \times 60\,\text{s}} = -1.79\,\frac{\text{m}}{\text{s}}$, where the "$-$" sign signifies westward.

★ Example 1.2: In Fig. 1.1(b), the displacement depends linearly on time, e.g. 0–2 s, 2–4 s, 4–8 s, 8–10 s, and 10–14 s time intervals, the instantaneous velocity is equal to average velocity in each time interval. Use Eq. (1.1) and calculate the velocities at times $t = 3$ s, 6 s, and 1 s.

Answer: Since the displacement vs time is linear in each time interval around $t = 3$ s, 6 s, and 1 s, the average velocity is also instantaneous velocity. Thus

a) The average velocity at $t = 3$ s is $\langle v \rangle = 0$ m/s, where $t_2 = 4$ s and $t_1 = 2$ s. The person is stationary.

b) The average velocity at $t = 6$ s is $\langle v \rangle = \frac{(4-(-4))}{(8-4)} = +2\frac{\text{m}}{\text{s}}$, where $t_2 = 8$ s and $t_1 = 4$ s.

c) The average velocity at $t = 1$ s is $\langle v \rangle = \frac{(-4-0)}{(2-0)} = -2\frac{\text{m}}{\text{s}}$, where $t_2 = 2$ s and $t_1 = 0$ s.

★ Example 1.3: Find the total walking distance at 1.0 m/s constant speed in 45 minutes.

Answer: For a given speed and time duration Δt, the distance traveled is (speed) × (time duration) or $d = v\Delta t = (1\ \text{m/s}) \times (45 \times 60\ \text{s}) = 2700$ m. There is no specific direction in this walk.

★ Example 1.4: In 5 hours, a car travels 250 miles on highway. What is the average speed?

Answer: The average speed is $v = \frac{250\,\text{mi}}{5\,\text{h}} = 50\frac{\text{mi}}{\text{h}}$.

★ Example 1.5: Convert (a) 340 m/s to mi/h; (b) 50 mi/h to m/s

Answer: (a) $340\frac{m}{s} = 340\frac{m}{s} \times \frac{1\,\text{mile}}{1609\,\text{m}} \times \frac{3600\,\text{s}}{1\,\text{hr}} = 760$ mile/hr.

(b) $50\frac{\text{mi}}{\text{hr}} = 50\frac{\text{mi}}{\text{hr}} \times \frac{1609\,\text{m}}{1\,\text{mile}} \times \frac{1\,\text{hr}}{3600\,\text{s}} = 22$ m/s.

3) Acceleration

Acceleration is the rate of velocity change, or the change of velocity per unit time:

$$a = \frac{\Delta v}{\Delta t} = \frac{v(t_2) - v(t_1)}{t_2 - t_1}. \tag{1.2}$$

The unit of acceleration is [Length/Time^2] or [m/s^2] in SI units. Figure 1.2 shows an example of velocity vs time (top) and the corresponding displacement with time (bottom) of an object from 0 to 14 s, where the velocity is constant in 5–8 s and 11–12 s, i.e. zero acceleration. At zero acceleration, the displacement is a straight line. At a positive constant acceleration, the displacement is an upward parabola. When acceleration is negative, the displacement curves downward.

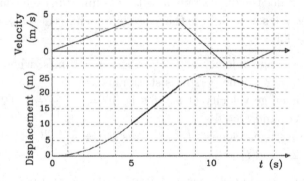

Figure 1.2: The top plot shows the velocity of an object from 0 to 14 s; the bottom plot shows the corresponding displacement in the 14 s interval.

★ Example 1.6: The Juno spacecraft arrived at Jupiter on July 4, 2016. It dives through intense barrages of radiation and reaching 130,000 miles per hour as it passed 2,900 miles above Jupiter's cloud tops. Express the speed of the Juno spacecraft in [m/s].

Answer: $v = 130{,}000 \left(\frac{\text{mi}}{\text{h}}\right) \times \left(\frac{1609\,\text{m}}{1\,\text{mi}}\right) \times \left(\frac{1\,\text{h}}{3600\,\text{s}}\right) = 58{,}000$ m/s

★ Example 1.7: Find the acceleration at time $t = 3$ s in Fig. 1.2. What is the displacement of the person in 14 s? What is the total distance that the person traveled?

Answer: In the time interval 1–5 s, the acceleration is $a = \frac{(4-0)\,\text{m/s}}{5\,\text{s}} = 0.8$ m/s^2; similarly, the acceleration is $a = 0$ m/s^2 in time interval 5–8 s, $a = \frac{(-2-4)\,\text{m/s}}{3\,\text{s}} = -2$ m/s^2 in time interval 8–11 s; $a = 0$ m/s^2 in time interval 11–12 s; and $a = \frac{(0-(-2))\,\text{m/s}}{2\,\text{s}} = +1$ m/s^2 in time interval 12–14 s. From the lower plot, we read the "displacement" is 21 m, while the total "distance" traveled is 26 m + 5 m = 31 m. The displacement is negative in the last 4 second from $t = 10$ s to 14 s.

★ Example 1.8: A car accelerates from 0 to 50 mi/hr in 10 s. The driver then applies the brake and stops in 4 s. What is the average acceleration in the first 10 s and then in the next 4 s, in m/s^2?

Answer: The accelerator for the first 10 second is

$$a = \frac{\left[(50-0)\frac{\text{mi}}{\text{hr}} \times \left(\frac{1609\,\text{m}}{\text{mi}}\right) \times \left(\frac{1\,\text{hr}}{3600\,\text{s}}\right)\right]}{10\,\text{s}} = 2.2\,[\text{m/s}^2].$$

The acceleration for the next 4 second is

$$a = \frac{\left[(0-50) \times \frac{1609}{1} \times \frac{1}{3600}\right]}{4} = -5.6\,[\text{m/s}^2].$$

C. Dynamics

Kinematics is the state of motion of particles. Dynamics deals with "why and how" objects achieve their acceleration or velocity. Isaac Newton, in his book *Mathematical Principles of Natural Philosophy*, laid the foundation of dynamics. Newton's 2nd law of motion states that the net force F is related to the mass m and acceleration a by

$$F = ma = m\frac{\Delta v}{\Delta t} = \frac{\Delta mv}{\Delta t} \equiv \text{"rate of momentum change."} \quad (1.3)$$

The acceleration is proportional to the net force acts on the object. The proportional constant is the "(inertial) mass." When the mass is in [kg] and acceleration in [m/s^2], the unit of force is [Newton] or [N], i.e. 1 [N] = 1 [kg · m/s^2] (see Appendix A for further discussions on Newton's three laws of motion).

★ Example 1.9: A batter hits a baseball, and the baseball changes momentum (mass × velocity) during the "contact time" between the baseball bat and the baseball. The average force on the baseball is the change in the momentum divided by the contact time between the ball and the baseball bat. The baseball bat experiences the same amount of force in the opposite direction (Newton's third law: action and reaction). A baseball with mass 0.145 kg moving at 40 m/s (90 mi/hr) is hit by a bat and moving exactly in the opposite direction at 50 m/s (112 mi/hr). If the contact time is 0.70 ms, what is the average force on the baseball?

Answer: We choose the positive direction as the direction outward, towards the field. The momentum of the incoming and outgoing baseball are $mv_i = -0.145 \times 40 = -5.8 \text{ kg} \cdot \text{m/s}$ and $mv_f = 0.145 \times 50 = 7.25 \text{ kg} \cdot \text{m/s}$, respectively. Using Eq. (1.3), we find the average force on the baseball as

$$F_{av} = \frac{mv_f - mv_i}{\Delta t} = \frac{[7.25 - (-5.8)] \text{ kg} \cdot \text{m/s}}{0.0007 \text{ s}} = 18{,}600 \text{ kg} \cdot \text{m/s}^2.$$

Some examples of force

1. Gravity: Newton discovered the universal law of gravitation, based on Kepler's laws on motion of heavenly bodies. The universal gravitational force between two point-objects with masses m_1 and m_2 at a distance r is

$$F = G\frac{m_1 m_2}{r^2}, \tag{1.4}$$

where $G = 6.67 \times 10^{-11}$ [N m^2/kg^2] is the universal gravitational constant. Weight is the Earth's gravitational force on a mass, i.e. Weight $= mg$ with $g = G\frac{M_E}{R_E^2} = 9.80 \text{ m/s}^2$, where $M_E = 5.98 \times 10^{24}$ kg and $R_E = 6.38 \times 10^6$ m are the mass and radius of the Earth. The gravitational force produces a free-fall acceleration $g = 9.80 \text{ m/s}^2$ toward the center of the Earth. In the US, we measure our weight in pounds [lb], or [1 lb = 4.45 N]. The "pound" is a force unit. It is equivalent to the weight of a 0.454-Kg mass, or 1 kg-weight is about 2.2 lb on Earth.

2. The "tension" is the force pulling along the string of string musical instruments, along cables of suspension bridges, etc.

3. The restoring force of a small perturbation to an elastic object in equilibrium is $F = -ky$, where k is the "spring constant," and y is the displacement from the equilibrium position. This is Hooke's law.

4. The air molecules inside a balloon are constantly bombarding and bouncing back from the balloon wall. The rate of momentum change of molecules produces a force that is balanced by the atmosphere pressure compressing on balloon. This force per unit area is "pressure."

Defining the *momentum* as mv, Newton's 2nd law in Eq. (1.3) shows that "the *net force is equal to the rate of change of momentum mv.*" This formulation of the 2nd law provides us a simple derivation of the wave speed in a medium to be discussed in Chapter 2.

III.   Pressure

A force can cause an object to move. For the disturbance of an extended object, the key physical quantity is the force perpendicular to a surface per unit area. We define the pressure as

$$p = F_\perp/A, \qquad (1.5)$$

where $F_\perp$ is the component of the force perpendicular to the surface of an object, and A is the surface area that the force applied on. The unit of pressure is $[N/m^2]$, or [Pascal] abbreviated with [Pa]. An old unit of pressure is $[dyne/cm^2]$, where 1 dyne $= 1$ g$\cdot$cm/s$^2 = 10^{-5}$ N with 1 $[N/m^2] = 10$ $[dyne/cm^2]$. Figure 1.3 shows the definition of pressure, examples of different pressure on the floor due to a high heel shoe and a running shoe, and a schematic of the bouncing of air molecules off container walls creating the air pressure inside a container. Just like in Example 1.9, the bouncing of air molecules in the container gives the air pressure on the container walls.

Figure 1.3: A schematic drawing of a force on a surface of area A. The pressure is equal to the applied force divided by the surface area. The pressure is higher when the contacting surface is smaller at a given magnitude of force. The pressure of a gas in a container arises from collisions and bouncing back of the gas molecules in the container.

Some Basic Information on Pressure

1. Based on Newton's second law of motion, the *force* on the wall of the container is equal to the *rate of change of particle momentum* hitting and bouncing back from the surface inside the container. One obtains the ideal gas law:

$$pV = NkT = nRT \,, \qquad (1.6)$$

where p is the gas pressure, $k = 1.38 \times 10^{-23}$ J/K is the Boltzmann constant, N is the number of molecules inside the volume V, $n = N/N_A$ is the number of moles, $N_A = 6.022 \times 10^{23}$ particles/mol is the Avogadro number, $T = 273 + T_C$ is the absolute temperature, T_C is the temperature in Celsius, and $R = N_A k = 8.314$ J/$^\circ$K is the gas constant.

2. The atmosphere pressure at sea level, $p_{\text{atm}} = 1$ atm $= 1.01325 \times 10^5$ [N/m^2] is literally equal to the weight of the column of air molecules in an area 1 m^2. Figure 1.4(a) shows the air pressure vs the altitude (height from Earth's surface). A unit often used in weather report is [mbar] (milli-bar) with 1 atm $= 1.01325$ Bar $= 1013.25$ mbar, where 1 bar $\equiv 10^5$ [Pa]. The air pressure at the center of hurricanes and tornadoes is substantially lower than the nominal atmospheric pressure. The lower the pressure, the stronger the hurricane. Similarly, the pressure under water increases about 1 atm for every 10 m of depth. The force on our chest, if we assume an average area of $A = 0.5$ m^2, is $F = p_{\text{atm}} A \approx 5 \times 10^4$ N or 11,000 pound. The air-pressure inside our chest pushes outward to balance the pressure from outside.

3. The measurement of subglottal pressure in human, or pressure in musical instruments often use the unit of [cm-H$_2$O], where 1 [cm-H$_2$O] = 98 [N/m^2] $\approx$ 100 [N/m^2].

4. Sound wave produces compression and rarefaction in a medium. When the medium is compressed, its pressure is higher, and when the medium is rarefied, its pressure is lower. The unit for pressure in sound waves is [Pascal] or [Pa]. The smallest sound pressure difference from atmospheric pressure that our ear can detect is $\Delta p_0 = 2 \times 10^{-5}$ Pa $= 20~\mu$Pa. A sudden loud sound pressure difference of $\Delta p_0 > 60$ N/m^2 can rupture our eardrums. Figure 1.4(b) shows the pressure waveform and spectrum of a recording of Beethoven's fifth opening phrase, obtained from Praat program: see http://www.fon.hum.uva.nl/praat/ (Ref. [1]).

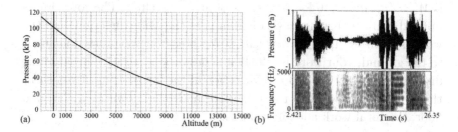

Figure 1.4: (a) The atmospheric air pressure vs the altitude on Earth's surface. (b) The sound waveform of the opening phrase of Beethoven's fifth symphony in Pascal and its spectrum.

★ Example 1.10: A 60 kg person stands on both feet. What is the pressure on the ground if she wears (a) a running shoe and (b) a high heel with contact areas 260 cm^2 and 26 cm^2 each respectively?

Answer: A 60 kg person has weight of (60 kg) × (9.8 m/s^2) = 588 N. At a standing position, if the surface area of each foot is 260 cm^2, the pressure is $588\,\mathrm{N}/(2 \times 0.026\,\mathrm{m}^2) = 11{,}300$ N/m^2. If the person wears high heels, where the contact surface is 26 cm^2, the pressure is 113,000 N/m^2.

IV.   Work, Energy, Power and Intensity

Waves transport wave energy and wave momentum as they propagate in the medium. This section discusses work, mechanical (kinetic and potential) energies, the work-energy principle, power and intensity.

A.   Work

The work is equal to the product of force and displacement in the *direction of the force*:

$$W = F \cdot x. \tag{1.7}$$

The unit of work is the Joule [J] $=$ [N $\cdot$ m]. If you carry a 10-kg box from the ground floor to the 10th floor at a height $h = 30$ m, the work is $W = (mg)h = (10 \text{ kg} \times 9.8\frac{m}{s^2}) \times 30$ m $= +2940$ J. The force mg is upward and the displacement of 30 m is in the same direction, and the work is positive. If you carry this 10-kg box from the 10th floor back to the ground, your work is -2940 J. The work is negative because the displacement is in the opposite direction of the force. Thus, if you carry the 10-kg box upward to the 10th floor and back down to the ground floor, the total work adds up to zero.

★ Example 1.11: (a) Find the work done in carrying a 2 kg computer up to the 3rd floor at 10-m height. (b) What is the work done for bringing the computer from the 3rd floor back to the 2nd floor at 5-m height? What is the total amount of work?

Answer: (a) The work is $W = [(2 \text{ kg}) \times (9.8 \text{ m/s}^2)] \times (10\text{m}) = 196$ J. (b) When bringing the computer downward, the force you hold the computer is upward $+(2 \times 9.8)$ N, and the displacement is downward -5 m. The work is $W = (2 \times 9.8) \times (-5) = -98$ J. This work is negative. The total work for carrying the computer to the 3rd floor and back down to the 2nd floor is $+196 - 98 = +98$ J.

On the other hand, if you carry 10-kg box and walk along a flat horizontal floor, you do *zero* work because the force of carrying the 10-kg box is vertical and the displacement is horizontal. Your displacement in the direction of force is zero, and thus the work is zero.

B.   Energy

Energy in physics is composed of many forms. In mechanics, there are kinetic energy and potential energy. In classical mechanics, the kinetic energy depends on the mass and velocity via

$$KE = \frac{1}{2}mv^2 . \tag{1.8}$$

The potential energy is the energy associated with the position of the particle. The unit of both the kinetic and potential energies is [Joule] = [J]. Some examples of potential energy are

a. The force on the Earth's *surface* is nearly uniformly pointing downward. Objects on Earth's surface have a potential energy of mgh, where m is the mass, $g = 9.80$ m/s^2 is the gravitational acceleration, and h is the height. Objects at higher altitude have a higher potential energy. They can fall from higher position and convert the potential energy into kinetic energy.

b. The potential energy of Newton's gravitational interaction of Eq. (1.4) is $V = -G\frac{m_1 m_2}{r}$. This is the energy holding our solar system together.

c. The restoring force of a small perturbing force on an object in equilibrium is $F = -ky$, where k is the spring constant, and y is the displacement from the equilibrium position. This is Hooke's law. The potential energy of Hooke's law is $PE = \frac{1}{2}ky^2$. The particle motion that obeys Hooke's law is "simple harmonic motion (SHM)."

A guitar string obeys Hooke's law. The potential energy of a guitar string, displaced by a small distance y at its mid-point, is $PE = \frac{2T}{L}y^2$, where T is the tension on the string, and L is the length of the string. The spring constant is $k = 4T/L$.

A bottle of gas of volume V at a pressure p_0 is a system in equilibrium. If the pressure is slightly perturbed by Δp, the restoring force obeys Hooke's law. The potential energy is $PE = \frac{V}{2p_0}(\Delta p)^2$.

★ Example 1.12: Find the kinetic energy of a baseball with a mass of 0.145 kg moving at 40 m/s.

Answer: The kinetic energy is $KE = \frac{1}{2}mv^2 = \frac{1}{2} \times 0.145 \times (40)^2 =$ 116 J.

★ Example 1.13: Find the potential energy of a guitar string 65 cm long with tension 55 N, displaced 8 mm at its midpoint.

Answer: The potential energy is $PE = \frac{2T}{L}y^2 = \frac{2\times55\,\mathrm{N}}{0.65\,\mathrm{m}}(0.008\ \mathrm{m})^2 = 0.0108$ J.

C. Work-Energy Principle and Conservation of Mechanical Energy

Kinetic energy and potential energy are "mechanical energy." The "work-energy principle" states that the mechanical energy at the initial state of a system plus work done on the system is equal to the mechanical energy at the final state of the system, i.e.

$$KE_i + PE_i + W = KE_f + PE_f. \tag{1.9a}$$

This is a statement of conservation of energy. If $W > 0$, the work is on the system; if $W < 0$, the work is by the system, e.g. friction force. If the work is zero, $W = 0$, the total mechanical energy is conserved, i.e.

$$KE + PE = \text{constant}. \tag{1.9b}$$

Many systems nearly obey the law of conservation of mechanical energy, e.g. the free fall of a ball a short distance, the pendulum motion of grandfather clocks, the vibration of simple harmonic oscillators, etc. In reality, a small friction force always exist to dissipate the mechanical energy.

★ Example 1.14: When a ball falls from $h = 10$ m height, what is its velocity at the bottom?

Answer: Applying the conservation of mechanical energy: $KE_i + PE_i = KE_f + PE_f$, where the initial kinetic energy is $KE_i = 0$, the initial potential energy is mgh, and the final potential energy is 0, we find $0 + mgh = \frac{1}{2}mv^2 + 0$. The speed of the ball at bottom is $v = \sqrt{2gh}$. The speed of the ball falling from 10 m height is $v = \sqrt{2 \times 9.8 \times 10} = 14$ m/s.

D. Power

The power in physics is the work or amount of energy delivered in a unit time,

$$\text{Power} = \frac{\text{Work}}{\text{time}} \quad \text{or} \quad \frac{\text{Energy}}{\text{time}}. \tag{1.10}$$

The unit of power is watt [W], where 1 [W] = 1 [Joule/s]. Another commonly used unit is the horse-power [hp] with 1 [hp] = 745.7 [W].

★ Example 1.15: Find the power in bringing a 2 kg computer up to 3rd floor at 10 m height in 20 s?

Answer: The power is $P = \frac{\text{Work}}{\text{time}} = \frac{2 \times 9.8 \times 10 \text{ J}}{20 \text{ s}} = 9.8$ W.

★ Example 1.16: A speaker using 10 W electric power with 0.5% efficiency. What is the acoustic power of the speaker? [(output power) = (input power) × (efficiency)]

Answer: The speaker delivers 10 × 0.5% = 0.05 W of sound power.

E. Intensity

The intensity in physics is the amount of power delivered or received per unit area:

$$\text{Intensity} = \frac{\text{Power}}{\text{Area}}. \tag{1.11}$$

The unit of intensity is [W/m^2]. The sound power we received through our ear canals is the intensity multiplied by the area opening of our ear canals. The intensity is the relevant physical quantity that our sound detection instruments can measure. Our hearing system can detect sound intensity at 10^{-12} W/m^2.

★ Example 1.17: What is the sound intensity at distance 10-m from the 0.2-W speaker?

Answer: We assume that the sound energy uniformly spreads over the entire spherical shell. The intensity is then $I = \text{Power/Area} = 0.2 \text{ W}/[4\pi(10 \text{ m})^2] = 1.6 \times 10^{-4}$ W/m^2, where the spherical surface area at radius $R = 10$ m is $4\pi R^2$.

★ Example 1.18: The solar constant is the solar energy per unit area measured above Earth's atmosphere. The solar constant is $1.36 \, \mathrm{kW/m^2}$. The Earth's radius is 6.37×10^6 m. Assuming an average 71% solar energy is absorbed, find the total solar power absorbed by the Earth.

Answer: The solar power reach the Earth is $P = \mathrm{Intensity} \times \mathrm{Area} \times \mathrm{Efficiency} = (1360 \, \mathrm{W/m^2}) \times (\pi R_{\mathrm{Earth}}^2) \times (71\%) = 1.2 \times 10^{17}$ W. Part of this solar energy is reflected back into the space. The Earth also radiates a smaller energy into space. The absorbed solar energy sustains all life on Earth.

V. Hooke's Law, Simple Harmonic Motion (SHM), Frequency and Period

Most objects around us do not change their shapes; they are in *"equilibrium."* When a small force (or *stress*) is applied to an object in equilibrium, a small deformation (or *strain*) is proportional to the applied stress. Removing the external force, the object will "restore" itself to its equilibrium shape. This restoring force obeys Hooke's law:

$$F = -ky\,, \tag{1.12}$$

where k is the *spring constant*, y represents the displacement from its equilibrium position, and the "negative sign" signifies the restoring nature of the elastic medium. The motion of an object obeying Hooke's law is simple harmonic, called simple harmonic motion (SHM).

The requirements for SHM are an equilibrium position and a linear restoring force. The position of SHM is given by a sine or cosine function at a single frequency. Most of the musical instruments are systems based on simple harmonic oscillation. Figure 1.5(a) shows a schematic drawing of a spring with spring constant k and a mass m hanging on the spring to reach an equilibrium position y_0. The net force is zero at equilibrium, i.e. $-ky_0 + mg = 0$. Making additional displacement of the mass, the mass will oscillate about the equilibrium position y_0. Figure 1.5(b) shows the measured displacement and velocity vs time. The time needed for one complete oscillation is a period T. The frequency is the number of oscillations per second, thus, $f = 1/T$. Figure 1.5(c) displays the same data for a longer time-period, showing the damping of a realistic SHM system. The period and frequency of the SHM for a system obeying Hooke's law are respectively

$$T = 2\pi\sqrt{\frac{m}{k}}, \quad \text{and } f \equiv \frac{1}{T} = \frac{1}{2\pi}\sqrt{\frac{k}{m}}, \tag{1.13}$$

where m is the mass, k is the spring constant, T is the period or the time needed to complete one oscillation, and f is the frequency or the number of oscillations per second.

The displacement from the equilibrium position is

$$y = A\sin(2\pi ft + \varphi) = A\sin\left(\frac{2\pi t}{T} + \varphi\right) \tag{1.14}$$

where A is the *amplitude* of oscillation, f is the *frequency*, T is the *period*, and φ is a phase angle depending on the initial condition (see Appendix B for a review of sinusoidal functions). For example, the SHM shown in Fig. 1.5(b) has $f = 1.4$ Hz, or $T = 0.73$ s, and $\varphi = 0$. The unit of frequency is Hertz [Hz], where 1 [Hz] $\equiv$ 1 [oscillation per second]. The frequency is independent of the amplitude in SHM. The velocity of the SHM is

$$u = 2\pi fA\cos(2\pi ft + \varphi). \tag{1.15a}$$

The velocity is always 90° $(\pi/2)$ ahead of the displacement. The velocity u is zero at the maximum displacement $|y|$ value; and the magnitude of

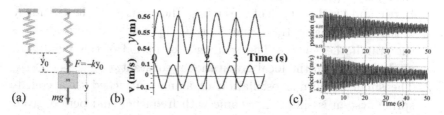

Figure 1.5: (a) A schematic drawing of a spring. When a mass hangs on the spring, the spring will extend its length to reach a new equilibrium at $mg = ky_0$. If the spring is further displaced to y from the equilibrium position, it will execute simple harmonic motion around the equilibrium position y_0. (b) Measured data of position and velocity vs time for a SHM oscillator showing the sine-cosine oscillation vs time. The variation of the amplitude corresponds to the systematic error of the measurement system. (c) The amplitude of oscillation decreases slowly in a weakly damped SHM oscillator.

velocity $|u|$ reaches its maximum at $|y| = 0$. The maximum velocity of SHM is related to the maximum displacement by

$$u_{\max} = (2\pi f)A = (2\pi f)y_{\max}. \qquad (1.15b)$$

A SHM system has a "natural" frequency given by Eq. (1.13). When the system is driving by a perturbing force, the amplitude of the SHM can become large if the frequency of the driving force is equal to the natural frequency of the SHM. This frequency is the *"resonant frequency"* of the SHM. A simple harmonic oscillator has only one resonance frequency. A few examples of nearly simple harmonic systems are the motion of simple pendulum, the oscillation of the automobile shock absorbers, the motion of string in string musical instruments, the motion of a piston in a tank, the oscillation of car shock absorbers, and the oscillation of a spring system, etc. We list some physical systems that nearly obey Hooke's law below.

1. A mass hung on a massless string and set to vibrate at small amplitude is a simple pendulum. The frequency and period of small amplitude oscillation are

$$f = \frac{1}{2\pi}\sqrt{\frac{g}{\ell}} \text{ and } T = 2\pi\sqrt{\frac{\ell}{g}}. \qquad (1.16)$$

Here g is the gravitational acceleration constant, or $g = 9.80 \text{ m/s}^2$ on Earth, and ℓ is the length of the string. The frequency is independent of the mass hung on the string. This is the basic design principle of the grandfather clocks. The frequency or period of this pendulum clock depends on the local gravitational constant g.

2. The small amplitude oscillation of a mass attached to a coiled-spring system is simple harmonic with frequency and period given by Eq. (1.13). We can use such a system to measure the mass of the attached object, even in weightless environments, e.g. a space station.

3. A tank of air is a Helmholtz resonator. Figure 1.6 shows a cylinder with area A and length L in a volume where air molecules are free to move. The spring constant is $k = \gamma pA/L$, where p is the air pressure and γ is the adiabatic gas constant. The resonance frequency of

Helmholtz resonator is

$$f = \frac{1}{2\pi}\sqrt{\frac{\gamma p A}{ML}} = \frac{v}{2\pi}\sqrt{\frac{A}{VL}}, \qquad (1.17)$$

where $M = \rho V$ is the mass of air in the resonator, ρ is the air density, V is the volume of the resonator, and $v = \sqrt{\gamma p/\rho}$ is the speed of sound wave in air (see Chapter 2). Applications of Helmholtz resonators are mufflers for noise reduction in engine exhausts, resonant bodies for musical instruments, etc.

Figure 1.6: (a) Shapes and sizes of Helmholtz resonators. Helmholtz used theses resonators to analyze frequency of a complicated sound source. (b) A schematic drawing of a Helmholtz resonator with neck cross-sectional area A, neck length L and volume V.

★ Example 1.19: When a mass of 2 kg is hung on a spring, the spring stretches 0.10 m [see Fig. 1.5(a)]. (a) What is the spring constant? (b) The oscillation frequency measured for an unknown object hung on this spring is 1.0 Hz. What is the mass of this object?

Answer:

(a) The spring stretch $y_0 = 0.10$ m to reach equilibrium with mass $m = 2$ kg. The gravity force mg is downward, and the restoring force $-ky_0$ is upward, the net force is zero in equilibrium, so $mg - ky_0 = 0$, or $k = \frac{mg}{y_0} = \frac{2.0\,\text{kg}\times 9.8\,\text{m/s}^2}{0.10\,\text{m}} = 196\frac{\text{N}}{\text{m}}$.

(b) From Eq. (1.13), the frequency is $f = \frac{1}{2\pi}\sqrt{\frac{196\,\text{N/m}}{m}} = 1.0$ Hz. We solve the equation to find the mass of the object as $m = 5.0$ kg. For a given spring constant, one can determine the mass of an object by the measurement of frequency.

★ Example 1.20: A grandfather clock uses a pendulum to keep its time at a 2-second period. What is the length of the pendulum?

Answer: The period of a pendulum is $T = 2\pi\sqrt{\frac{\ell}{g}}$, or $\ell = g(\frac{T}{2\pi})^2 =$ $9.8\frac{m}{s^2} \times (\frac{2s}{2\pi})^2 = 0.9929$ m. The period of the pendulum depends on the local gravitational constant g. Fortunately, we can adjust the length of the pendulum to adjust the timepiece. If we bring this pendulum clock to the Moon, where g is smaller than that on Earth, the period of the pendulum is longer and runs slower. Similarly, the natural frequency in our walking pace depends on the length of our legs.

★ Example 1.21: A small flask consists of a sphere 9.8 cm in diameter plus a cylindrical pipe with a neck 3 cm in diameter and 10 cm in length. What is the resonance frequency?

Answer: The area of neck is $A = \pi r^2 = 0.000707$ m^2, where $r = 0.015$ m. The flask is composed of a sphere and a cylinder. The volume is $V = (4\pi/3)R^3 + AL = 0.000500$ m^3, where $R = 0.049$ m is radius of the sphere and $L = 0.10$ m. Using Eq. (1.17) with the sound speed $v = 343$ m/s, we find the resonance frequency $f = \frac{v}{2\pi}\sqrt{\frac{A}{VL}} = \frac{343\,\text{m/s}}{6.28}\sqrt{\frac{0.000707\,\text{m}^2}{0.000500\,\text{m}^3 \times 0.010\,\text{m}}} = 649$ Hz.

A. Kinetic and Potential Energies and Damping of the Vibrational Motion

The kinetic energy and potential energy of a simple harmonic motion are respectively

$$KE = \frac{1}{2}mu^2 \text{ and } PE = \frac{1}{2}ky^2, \tag{1.18a}$$

where u is the velocity of the mass and y is the displacement of the mass. For a system without friction force, the mechanical energy is conserved, i.e.

$$E = KE + PE = \frac{1}{2}mu^2 + \frac{1}{2}ky^2 = \frac{1}{2}mu_{max}^2 = \frac{1}{2}ky_{max}^2, \tag{1.18b}$$

where u_{max} and y_{max} are the maximum speed and maximum displacement [see Eq. (1.15b)]. Any real oscillator has friction force that dissipates energy. The amplitude of the dissipative SHM gradually decreases, shown in Fig. 1.5(c).

B. Simple Harmonic Wave

A wave is a disturbance that travels in space or a medium. All waves transport information or energy and momentum from one position to another. Simple harmonic motion in a medium forms a simple harmonic (SH) wave. For example, the simple harmonic motion of air-molecules (forward/backward) forms a simple harmonic sound wave in air, and simple harmonic motion of a piano-string (up/down) forms a simple harmonic wave travels along the string. The equations of displacement of the medium by a simple harmonic wave of a frequency f (or period T) and wavelength λ traveling to $+x$ direction and $-x$ direction are respectively

$$y = A\sin\left(\frac{2\pi x}{\lambda} - \frac{2\pi t}{T} + \varphi_0\right) = A\sin\left(\frac{2\pi x}{\lambda} - 2\pi ft + \varphi_0\right), \quad (1.19a)$$

$$y = A\sin\left(\frac{2\pi x}{\lambda} + \frac{2\pi t}{T} + \varphi_0\right) = A\sin\left(\frac{2\pi x}{\lambda} + 2\pi ft + \varphi_0\right). \quad (1.19b)$$

where A is the amplitude of the wave, λ is the wavelength, T is the period, f is the frequency, and φ_0 is an arbitrary phase. The wavelength λ is the distance from the crest to crest. At a position x, the medium (or particle) oscillates in SHM. The combination of the SHM of all particles produces a SHM wave that carries wave energy in the wave propagation direction. The maximum *particle* speed is $u_{max} = (2\pi f)y_{max}$ in Eq. (1.15b) while the *wave* speed is

$$v = f\lambda = \lambda/T. \quad (1.20)$$

Figure 1.7 shows schematically the wave $y(x,t)$ at various time $t = 0$, 2, 4, 6, 8 offset by 2-units intervals with period $T = 8$ and wavelength $\lambda = 1$ with arbitrary units. If the units of wavelength and time are [meter] and [millisecond], respectively, then the wavelength is $\lambda = 1$ m, and wave frequency is $f = 1/(8 \text{ ms}) = 125$ Hz. The wave speed is $v = f\lambda = 125$ m/s. The offset provides easier viewing of the wave propagation with time.

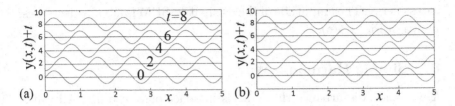

Figure 1.7: Schematic drawing of SH waves traveling in the $+x$ direction (a) and in the $-x$ direction (b) at $\varphi_0 = 0$. Each trace represents the wave in space at a given time. The wavelength is the distance that the wave repeats in space and the period is the time that the wave repeats itself. These waves travel visibly toward $+x$ and $-x$ directions, respectively.

★ Example 1.22: A simple harmonic wave given by $y = (-0.5\,\text{m}) \sin (5\pi x - 2000\pi t)$ travels in the $+x$ direction, where x is in [m], and t in [s]. What are the amplitude, wavelength, frequency, and the wave speed?

Answer: Since $(-0.5\text{ m}) \sin(5\pi x - 2000\pi t) = (0.5\text{ m}) \sin(5\pi x - 2000\pi t + \pi)$, we identify the wave equation to Eq. (1.19a) and find $A = 0.5$ m; $\frac{2\pi x}{\lambda} = 5\pi x$ or $\lambda = 0.4$ m; $2\pi f t = 2000\pi t$ or $f = 1000$ Hz; and $\varphi_0 = \pi$ rad. Thus $v = f\lambda = 400$ m/s. This is a simple harmonic wave traveling in the $+x$ direction. The phase φ_0 determines the wave disturbance at $t = 0$ and $x = 0$.

★ Example 1.23: A simple harmonic wave $y = (0.1\text{ m}) \sin(\frac{2\pi x}{0.75} + \frac{2\pi t}{0.01} + \frac{\pi}{2})$ travels in the $-x$ direction, where x is in [m] and t in [s]. Find the amplitude, wavelength, frequency and wave speed.

Answer: Identifying the wave equation in Eq. (1.19b) that describes a wave traveling in the $-x$ direction, we find $A = 0.5$ m; $\frac{2\pi x}{\lambda} = \frac{2\pi x}{0.75}$ or $\lambda = 0.75$ m; $\frac{2\pi t}{T} = \frac{2\pi t}{0.01}$ or $T = 0.01$ s, so $f = 100$ Hz; and $\varphi_0 = \frac{\pi}{2}$ rad. Thus, the wave speed is $v = f\lambda = 75$ m/s.

Figure 1.8 shows the definition of a waveform for transverse and longitudinal sinusoidal waves at an instant of time in space. (a) When the motion of the medium (particle) is perpendicular to the direction of the wave propagation, the wave is a transverse wave; (b) When the motion of the medium is parallel to the direction of wave propagation, it is a longitudinal wave.

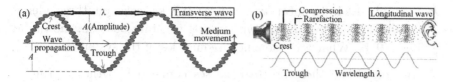

Figure 1.8: (a) A transverse sinusoidal wave travels to the right at an instant of time. (b) A sinusoidal longitudinal wave at an instant of time. The amplitude is the difference between the maximum pressure and the average pressure.

The wave crest is the maximum displacement, and the wave trough is the minimum displacement. The distance of a complete wave oscillation is called wavelength λ, i.e. the wavelength is the distance crest to crest or trough to trough. The maximum displacement is the wave amplitude A. Particles in the medium move only a small displacement around the equilibrium position, while the wave energy and momentum travel forward at the wave speed. The wave does not transport particle in the medium. For a longitudinal wave in a medium, molecules move forward and backward causing increase or decrease in pressure. The wave, carrying energy and momentum, travels forward in space; while air molecules oscillate only locally.

VI. Modes of Oscillations (Standing Waves and Resonances)

When a single mass hangs on a massless-spring, there is only one resonance frequency given by Eq. (1.13). With two masses connected with massless springs, there are two normal modes at frequencies f_a and f_b, shown in Fig. 1.9. These two normal modes are resonances of the system.

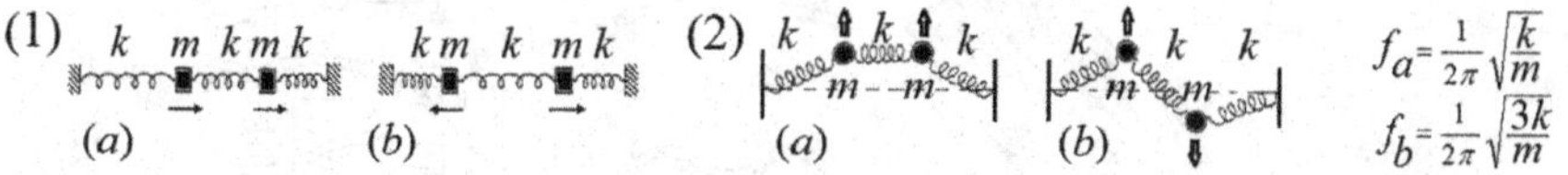

$$f_a = \frac{1}{2\pi}\sqrt{\frac{k}{m}}$$

$$f_b = \frac{1}{2\pi}\sqrt{\frac{3k}{m}}$$

Figure 1.9: Schematic plot of 2 masses coupled through massless-springs in longitudinal motion (1) and transverse motion (2). There are two normal modes, a and b. Two masses oscillate together in mode a, and oscillate opposite to each other in mode b. Their mode frequencies are f_a and f_b, respectively.

★ Example 1.24: Connect two masses 0.5 kg each on three identical ideal massless-springs with spring constant 50 N/m, shown in Fig. 1.9(1). What are resonance frequencies of the system?

Answer: When there are two masses connected to these springs, there are two normal modes at frequencies

$$f_a = \frac{1}{2\pi}\sqrt{\frac{50}{0.5}} = 1.59 \text{ Hz}; \quad \text{and} \quad f_b = \frac{1}{2\pi}\sqrt{\frac{3\times50}{0.5}} = 2.76 \text{ Hz}.$$

A system with three masses connected to four massless springs has three normal modes; the system with 10 masses distributed on a massless spring has 10 normal-modes, and so on. A string, a rod, or a membrane has uniform mass distribution; there are *infinite* number of modes. The resonance frequencies depend on its geometry and boundary conditions of a physical system. The lowest frequency of these modes is the fundamental frequency.

The wave produced by some physical systems, such as a string or a cylindrical pipe, are periodic. Frequencies of higher order modes are integer multiples of a fundamental frequency. These higher order modes are "*harmonics*." However, waves produced by many complex vibrational systems are *not periodic*, their mode frequencies are *not* integer multiples of the fundamental frequency. The term "*overtones*" denote all resonances with frequencies higher than the fundamental mode frequency. The term "*partials*" refers to all modes of vibration of a system including the fundamental mode and all overtones. The term "*upper partials*" is equivalent to "overtones." Figure 1.10 shows examples of some resonant modes of a string, a membrane, and a rod (see Ref. [24]).

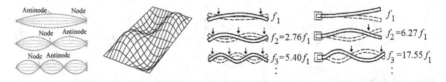

Figure 1.10: Schematic drawing of some low frequency standing waves of a string, a membrane, and uniform bars. Since these systems have uniform mass distribution, there are infinite number of modes.

VII. Spectra of Musical Instruments and Complex Waves

A complex wave is a superposition of simple harmonics waves with many frequencies. Waves produced by all musical instruments are superposition of many harmonics. The "spectrum" is the frequency content of a vibrational system. Fourier analysis (see Appendix B) is a powerful mathematical theory to extract these frequencies. The digital Fast Fourier Transform (FFT) algorithm is a powerful computer algorithm that can efficiently extract amplitudes and frequencies of a complex wave. The "spectrum analyzer" is the instrument that extracts spectrum of complex waves in real-time.

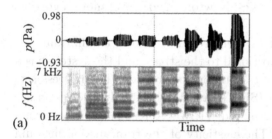 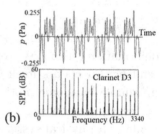

Figure 1.11: (a) Sound pulse (top) and spectrum (bottom) of a flute playing the diatonic scale. (b) The waveform of a clarinet playing D3 note at 146.8 Hz (top), and its spectrum (bottom). Since these waveforms are periodic, all higher-order modes are integer multiples of the fundamental frequency. The air column is a uniform "continuous" medium; there are infinite number of modes.

A *periodic* wave is a wave that repeats after a fixed time-interval. A periodic wave is made of sinusoidal waves at frequencies that are integer multiples of the fundamental frequency, called *harmonics*. The term "harmonics" may sometimes contain both the fundamental and higher modes. Figure 1.11 shows examples of sound waveforms: (a) a flute playing a diatonic scale and (b) a clarinet playing the D3 note at 146.8 Hz. The top plots show pressure waveform vs time, while lower plots show their spectra. The lowest frequency mode is the fundamental and all higher frequency modes are harmonics.

The envelope of the flute diatonic scale waveform in Fig. 1.11(a) is characteristic of the flute sound. If one zooms in on each note, the waveform is nearly periodic. The fundamental mode frequency is $f_0 = 1/T_0$, where T_0 is the period of the periodic wave. Its spectrum is composed of a fundamental mode and harmonics that are integer multiple of the fundamental frequency. The waveform of the clarinet in Fig. 1.11(b) is clearly periodic and its spectrum shows that higher-order modes are integer multiples of the fundamental at 146.8 Hz.

VIII.   Summary

Waves are the disturbance of physical quantities that propagate in the space and time. All waves transport energy and momentum along the wave's propagation path. Sound wave is a pressure disturbance that travels in an elastic medium. The motion of the medium for wave propagation obeys classical mechanics. Newton's 2nd law of motion governs the particle motion.

When a disturbance (stress) perturbs a system in equilibrium, the deformation (strain) is proportional to the stress, and the system has a tendency to *restore* itself to its equilibrium position. The process sets up simple harmonic motion (SHM). A SHM have only one resonance at resonant frequency given by Eq. (1.13). A resonance is a normal mode of a physical system. The elasticity of the transmission medium in equilibrium provides the oscillatory motion in waves. Each particle in the medium oscillates about its equilibrium position, while the wave carries the wave energy and wave momentum forward. The combination of simple harmonic motion of all particles forms a simple harmonic wave. A simple harmonic wave has a wave frequency f and wavelength λ. The wave speed is $v = f\lambda$. We call wave "speed" instead of wave "velocity" because a wave can propagate in all directions from a source.

A realistic physical system may have infinite normal modes. The frequencies of these modes depend on the intrinsic properties of the medium such as density, mass, shape, geometry and boundary condition. When the frequency of an external driving force is equal to one of these mode frequencies, the perturbing becomes large. Each particle in the medium oscillates about its equilibrium position, while the wave carries the wave energy and momentum forward. Using classical mechanics, we can link wave energy and momentum to the energy and momentum of the particle motion. The work energy principle of

Eq. (1.19) of classical mechanics provides the basis of energy conservation in wave transmission.

A complex wave is a superposition of many simple harmonic waves. We can use the Fast Fourier Transform [FFT, see Appendix B(2)] algorithm to decompose waves into components of simple harmonic waves. The frequency of the wave is the same as the frequency of particle oscillation in the medium. The "spectrum" is the frequency content of a sound. The lowest frequency of the spectrum is the "fundamental." Higher frequency components are "overtones." If higher harmonics are integer multiples of the fundamental, the wave is periodic. The fundamental frequency is the inverse of the period. Most musical instruments, except some percussion instruments, have periodic wave structure. Their spectrum is harmonic, such as shown in Fig. 1.11. Human speech sounds come from the periodic air pulses produced by the opening and closing of the vocal folds. The frequency of the periodic air pulses is the "pitch" of human speech and the singing voice.

We generally classify waves into the longitudinal or transverse wave. Sound wave is a longitudinal wave, where particles in the medium oscillate forward and backward in the propagation direction of sound wave. This process compresses and rarefies the medium, and produces pressure disturbance that travels in the medium. The wave energy in the medium is the sum of particle kinetic and potential energies. Using classical mechanics, we can link wave energy and wave momentum to the energy and momentum of the particle motion in medium (see Chapter 2).

IX.   Homework 01

1. Your car has gas mileage of 35 miles per gallon or 15 km/liter in SI units (1 gallon = 3.78 liters). (a) At a cruising speed of 65 miles per hour, what is the cruising speed in SI units = _____ m/s. (b) The number of gallons of gasoline needed to travel 350 miles is _____ gallon. (c) The time (t) required to cover 350 miles _____ hours.

2. The graph below shows the position x of an object versus the time t.

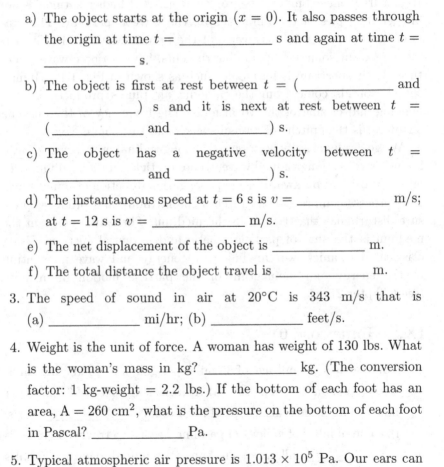

a) The object starts at the origin ($x = 0$). It also passes through the origin at time $t =$ _____ s and again at time $t =$ _____ s.

b) The object is first at rest between $t =$ (_____ and _____) s and it is next at rest between $t =$ (_____ and _____) s.

c) The object has a negative velocity between $t =$ (_____ and _____) s.

d) The instantaneous speed at $t = 6$ s is $v =$ _____ m/s; at $t = 12$ s is $v =$ _____ m/s.

e) The net displacement of the object is _____ m.

f) The total distance the object travel is _____ m.

3. The speed of sound in air at 20°C is 343 m/s that is (a) _____ mi/hr; (b) _____ feet/s.

4. Weight is the unit of force. A woman has weight of 130 lbs. What is the woman's mass in kg? _____ kg. (The conversion factor: 1 kg-weight = 2.2 lbs.) If the bottom of each foot has an area, A = 260 cm^2, what is the pressure on the bottom of each foot in Pascal? _____ Pa.

5. Typical atmospheric air pressure is 1.013×10^5 Pa. Our ears can detect the pressure wave of a sound wave at 20 μPa. What fraction of the threshold sound pressure to the atmospheric pressure? Ratio = _____ .

6. The temperature outside in the spring is 70°F, the temperature in Celsius scale is _____ °C (see Appendix D for unit conversion formula).

7. The energy required to lift an 8.0-kg lead brick to a height of 2 meters is _____ J. If you do the lifting in 1.5 s, the average power is _____ watts. (The Earth's gravitational acceleration constant is $g = 9.80$ m/s^2.)

8. (a) A baseball with a mass of 0.145 kg is pitched at 40 m/s. The kinetic energy of the baseball is _____ J. (b) The baseball is brought to a vertical height of 81.6 m above ground. The potential energy of the baseball is _____ J.

9. The wave speed of all wave is related to the frequency and wavelength by $v = f\lambda$. The EM wave can be classified into radio wave; visible-light ($\lambda = 400$–700 nm), X-ray; and γ-ray according to the wave frequency. The speed of EM wave in vacuum is 299792458 m/s exactly, or approximately 3×10^8 m/s. An FM radio broadcasting at 104.7 MHz frequency, what is the wavelength of the radio wave? $\lambda = $ _____ m.

10. The wave speed of all wave is related to the frequency and wavelength by $v = f\lambda$. The speed of sound in air at 20°C is 343 m/s. Ears of young men can hear sound from 20 Hz to 20,000 Hz. The wavelength of the sound wave at 20 Hz is _____ m, and the wavelength of sound wave at 20,000 Hz is _____ m.

11. In simple harmonic motion (SHM), the relationship between the "restoring" force, F, and the displacement, y, from the "equilibrium" position is (linear/nonlinear) (choose one), i.e. the requirements for SHM are an _____ position and a linear _____ force.

12. The graph below shows displacement y and speed v versus time for simple harmonic motion, with specific times t_1, t_2, t_3, t_4 and t_5.

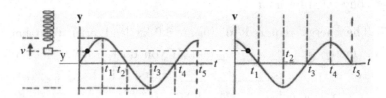

(a) The speed reaches its maximum value when displacement is zero at times _____ , and _____ . (b) The period of this SHM is _____ . (c) The maximum distance from the equilibrium position, when the speed of the particle is zero, occurs at _____ , _____ , and _____ .

13. The phenomenon that occurs when the frequency of the driving force is the same as the "natural" or "normal" frequency of the oscillating system is called _____ .

14. A mass of 0.50 kg hanging vertically on a particular spring causes its length to increase by 0.25 m. (a) The spring constant is $k = $ _____ [kg/s^2]. (b) When the system set up SHM around its equilibrium position with an amplitude 0.10 m, its frequency is _____ [Hz]. (c) The maximum speed of the SHM is _____ m/s. The total energy of this SHM is _____ J.

15. A pendulum has a length of 2.0 m, located on Earth with $g = $ 9.80 m/s^2. For small angle oscillations, the frequency is $f = $ _____ Hz and the period $T = $ _____ s.

16. In vibrating systems, the motion repeats in a time interval called the _____ , T. The _____ , $f = 1/T$, has units of Hertz (1 Hz $=$ 1 oscillation/second).

17. A grandfather clock runs slow. You should move the bob (<u>farther away</u>/<u>nearer to</u>) (choose one) the supporting point to adjust the clock time.

18. In a simple harmonic wave, each particle in the medium is oscillating with _____ motion. The frequency of the wave is the _____ of the oscillating particle; the amplitude of the wave is the _____ of the oscillating particle.

19. A wave that repeats in a fixed period is a _____ wave. A periodic wave can be decomposed into frequencies that are integer multiples of each other. The lowest frequency mode of the wave is called the _____ mode, and the frequencies of higher modes having integer multiples of the fundamental frequency are

    _____ .

20. A simple harmonic wave traveling to the $+x$ direction is $y = (0.10\,\text{m})\sin\left(\frac{2\pi x}{2.0} - \frac{2\pi t}{0.01}\right)$, where x is in [m], and t in [s]. The amplitude of this wave is _____ m; the wavelength of this wave is _____ m; the frequency of this wave is _____ Hz; the maximum velocity of the motion of the medium is _____ m/s, and the wave speed is _____ m/s.

21. A simple harmonic wave traveling to the $-x$ direction is $y = (0.10\,\text{m})\sin\left(\frac{2\pi x}{2.0} + \frac{2\pi t}{0.01}\right)$, where x is measured in [m], and t in [s]. The amplitude of this wave is _____ m; the wavelength of this wave is _____ m; the frequency of this wave is _____ Hz; the maximum speed of the motion of the medium is _____ m/s, and the wave speed is _____ m/s.

22. A complex wave is a superposition of simple harmonics waves with many _____ .

23. The frequency limits of human hearing are about 20 Hz to 20 kHz. The periods of these oscillations are _____ ms and _____ μs, respectively. The corresponding wavelengths of the sound wave in air at 20°C are _____ m and _____ m, respectively. (Note that the speed of sound wave in air at 20°C is 343 m/s.)

24. If the wave speed is independent of its frequency, it is a "non-dispersive" wave. The wavelength is the wave _____ divided by the _____ . A non-dispersive complex wave is made of two frequencies, 100 Hz and 300 Hz. The wave speed is 240 m/s. The wavelengths of the 100 Hz and 300 Hz components are respectively _____ m and _____ m.

25. The frequencies of oscillations with periods: 0.2 s, 10 ms, and 20 μs are respectively _____ Hz, _____ Hz, _____ kHz.

26. The periods of oscillations with the following frequencies, 20 Hz, 5000 Hz, 10 kHz are _____ s, _____ ms, _____ μs.

27. You stand 80 m away from a cliff and loudly yell "hello." How long does it take you to hear the echo? _____ s. (Assume speed of sound is 343 m/s, and remember the sound has to travel there and back.)

28. Electromagnetic waves travel through space at a speed of 3×10^8 m/s. The frequency of microwaves with $\lambda = 3$ cm (used in police radar) is _____ Hz.

29. A *periodic* wave is a wave that repeats in a fixed period. We can decompose a periodic wave into sinusoidal waves at frequencies that are integer multiples of the fundamental frequency, called *harmonics*. A periodic wave has a period of 8.0 ms, what

is the frequencies of its first 4 harmonics: _____ Hz,
_____ Hz, _____ Hz, and _____ Hz.

30. When waves travel, we call *wave speed* instead of *wave velocity* because they normally travel in *all* directions from a wave source. A church bell is rung at the midnight on the New Year eve. The time for a person to hear the new-year bell at the east side of town 10 km from the bell tower is _____. The time for a person to hear the new-year bell at the south of the town 10 km from the bell tower is _____. (The speed of the sound wave is 331 m/s in all directions.)

Chapter 2

Wave, Standing Waves, Wave Intensity and Impedance

A wave is a disturbance that propagates in a medium. It transports wave energy, wave momentum and information from one space-time location to another. Examples of waves are sound waves, earthquake waves, water waves, electromagnetic waves (including radio, microwave, millimeter-wave, visible light, X-ray, gamma ray), etc. An electromagnetic wave is the propagation of oscillatory electric and magnetic fields that can propagate in vacuum at speed $c \equiv 299792458$ m/s. Sound waves cannot propagate in vacuum, because there is no medium to transmit the pressure disturbance.

We loosely classify waves into longitudinal and transverse waves. The medium in a transverse wave oscillates perpendicular to the direction of wave propagation. Examples of transverse waves are wave propagation on a string, on a membrane, etc. An electromagnetic wave is a transverse wave because the electric and magnetic fields oscillate perpendicular to the EM wave propagation direction. When the disturbance causes the medium to oscillate along the wave propagation direction, the wave is a longitudinal wave. Examples of longitudinal waves are sound pressure waves, earthquake P-waves, etc. Water waves on the ocean's surface are a combination of longitudinal and transverse waves. This chapter discusses wave properties, such as *wave speed, wave energy and intensity; Huygens' principle; reflection; refraction; diffraction; superposition, interference and beat; inverse square law; and the Doppler effect.*

I.   Wave Properties

We classify waves into longitudinal and transverse waves, shown in Fig. 2.1(a). The medium of a longitudinal wave vibrates along the direction of wave propagation, e.g. the forward and backward motion of the slinky. The vibration of the medium of a transverse wave is perpendicular to the direction of wave propagation, e.g. the up/down motion of a rope. Figure 2.1(b) defines the wave amplitude and phase, where two sine waves have a 90° or $\pi/2$ in phase-shift. A wave can have an offset. In electronics, one can generate a wave from a function generator with a DC bias or DC offset. A sound wave in air is a small pressure perturbation around the atmospheric pressure p_{atm}.

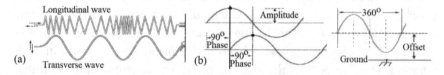

Figure 2.1: (a) Models of the longitudinal and transverse waves. The medium in a longitudinal wave oscillates in the same direction of the wave motion, while the medium of transverse wave oscillates in the direction perpendicular to the wave propagation. (b) The phases of two sine waves $A \sin\left(\frac{2\pi x}{\lambda} - \frac{2\pi t}{T} + \frac{\pi}{2}\right)$ and $A \sin\left(\frac{2\pi x}{\lambda} - \frac{2\pi t}{T}\right)$ differ by a phase of 90° or $\frac{\pi}{2}$ rad. A wave can have a constant offset to a ground level. For example, the sound wave is a small oscillation offset on top of the atmospheric pressure $p_{\text{atm}} = 1.013 \times 10^5$ Pa.

A wave can be periodic or non-periodic, continuous or pulsed. According to the Fourier theorem (see the Appendix B), all waves can be decomposed into components of sinusoidal waves, characterized by amplitudes, wavelengths, frequencies, and phases. A sound wave is a longitudinal pressure wave. Important characteristics of the sound wave to human ears are *loudness*, *pitch* and *timbre*, related to wave amplitude, wave frequency content, and the pulse shape of a sound.

A.   Speed of the Transverse Wave on a Uniform String

We consider a transverse wave traveling at the wave speed v on a uniform string, where the mass per unit length string is $\mu = M/L$ with the mass M and the length L. Consider a small section of the string Δx, where the mass is $\mu \Delta x$. As the wave travels, the wave momentum is equal to mass times wave velocity, i.e. $(\mu \Delta x)v$. The time for the wave to travel a distance Δx of this section is $\frac{\Delta x}{v}$. Applying Newton's 2nd law, which states that the net force on the wave is equal to the rate of change of wave momentum, we find that the "tension" on the string is $F_T = \frac{\Delta(\text{momentum})}{\Delta(\text{time})} = \frac{[(\mu \Delta x)v]}{\Delta x/v} = \mu v^2$, or

$$v = \sqrt{\frac{F_T}{\mu}}. \tag{2.1}$$

The *kinetic energy* per unit length is $\frac{1}{2}\mu u^2$, where u is the speed of the medium in up-down oscillations. The *total energy* per unit length is $\frac{1}{2}\mu u_{\max}^2$, where $u_{\max} = 2\pi f y_{\max}$ is the maximum speed of the medium [see Eq. (1.15b)], f is the frequency, and $y_{\max}$ is the maximum displacement of the transverse vibrational motion of the medium (see Chapter 1). The *potential energy* is equal to the total energy minus the kinetic energy.

B.   Speed of the Longitudinal Wave in a Uniform Medium

The medium in a longitudinal wave oscillates forward and backward along the wave's propagation direction. However, a wave does not transport "mass" as it propagates; rather, it transports wave energy and momentum forward. Consider a longitudinal plane wave traveling in the x-direction with velocity v through a slab of area A and length ℓ. The volume of the slab is $V = A\ell$.

When a net force F is applied to the medium in an area A, the pressure is $\Delta p = F/A$ and the volume change is $\Delta V = A \cdot \delta x$, where δx is the compression or expansion due to the wave pressure. If the pressure increases, the change in the volume is negative, or vice versa. Since wave pressure $\Delta p > 0$, and $\Delta V < 0$, the mass associated with the perturbed volume is $-\rho \Delta V$, where the negative sign signifies the positive mass. The wave momentum is $[-\rho \Delta V]v$. The net force F is rate of change of the wave momentum through the slab, or $[-\rho \Delta V]v/\Delta t$, where $\Delta t = \ell/v$ is the wave traveling time through the slab, or $F = \frac{(-\rho \Delta V)v}{\ell/v}$. The stress is

$$\Delta p \equiv \frac{F}{A} = -\rho v^2 \frac{\Delta V}{A\ell} = -\rho v^2 \frac{\Delta V}{V} \equiv -B\frac{\Delta V}{V}. \tag{2.2}$$

Here we apply the law of elasticity of the medium: $\frac{F}{A} \equiv -B\frac{\Delta V}{V}$, where B is the elastic modulus or bulk modulus of the medium [see Appendix A(3)]. Neglecting a small correction factor of the elastic modulus in solids, the speed of a longitudinal wave in the medium becomes

$$v^2 = \frac{B}{\rho}, \quad \text{or} \quad v = \sqrt{\frac{B}{\rho}}. \tag{2.3}$$

Since the elastic modulus and density are intrinsic property of the medium, the speed of wave depends only on the intrinsic properties of the medium. As wave propagates in medium, each atom make simple harmonic motion around its equilibrium position with spring constant $k = BA^2/V$.

The potential energy of the slab is $\frac{1}{2}k \cdot \delta x^2 = \frac{1}{2}\frac{F^2}{k} = V\frac{1}{2B}(\frac{F}{A})^2 = V\frac{1}{2\rho v^2}\Delta p^2$, where $F = -k \cdot \delta x$ and $\Delta p = \frac{F}{A}$. Thus the wave potential energy per unit volume is $\frac{1}{2\rho v^2}\Delta p^2$. The kinetic energy per unit volume is $\frac{1}{2}\rho u^2$, where u is the velocity of the molecules in wave motion. The total "wave energy density" (or wave energy per unit volume) is the sum of kinetic and potential energies:

$$w = \frac{1}{2}\rho u^2 + \frac{1}{2\rho v^2}\Delta p^2 = \frac{1}{2}\rho u_{\text{max}}^2 = \frac{1}{2\rho v^2}\Delta p_{\text{max}}^2. \tag{2.4}$$

Note that u and Δp are conjugate variables with $\Delta p_{\text{max}} = \rho v u_{\text{max}}$. The intensity of a wave is

$$I = wv = \frac{1}{\rho v}\frac{1}{2}\Delta p_{\text{max}}^2 = \frac{1}{\rho v}\Delta p_{\text{rms}}^2. \tag{2.5}$$

Here we use the fact that $\Delta p_{\text{max}} = \sqrt{2}\Delta p_{\text{rms}}$ [see Appendix B(4)].

★ Example 2.1: What is longitudinal wave speed in a bulk of steel? (The density and the elastic modulus of steel are $\rho = 7700$ kg/m^3 and $B = 19.5 \times 10^{10}$ N/m^2 respectively).

Answer: The speed of the longitudinal wave in steel is $v = \sqrt{\frac{B}{\rho}} = \sqrt{\frac{19.5 \times 10^{10}\,\text{N/m}^2}{7700\,\text{kg/m}^3}} = 5000\frac{\text{m}}{\text{s}}$.

★ Example 2.2: What would the tension of a steel wire be if the speed of the transverse wave of the steel wire at diameter 1 mm were equal to the speed of the longitudinal wave in Example 2.1?

Answer: The diameter of the steel string is 1 mm and its radius is $r = 0.0005$ m. The volume of a uniform string of length L is $\pi r^2 L$, the mass per unit length is

$$\mu = \frac{\rho(\pi r^2 L)}{L} = \rho \pi r^2 = \left(7700 \frac{\text{kg}}{\text{m}^3}\right) \times (3.14)(0.0005\text{m})^2 = 0.00604 \frac{\text{kg}}{\text{m}}.$$

Using Eq. (2.1), we find that the tension needed to achieve the wave speed of 5000 m/s is

$$F_T = \mu v^2 = (0.00604)(5000)^2 = 150,000 \text{ N}.$$

The "stress" on the wire is $F_T/(\pi r^2) = 1.9 \times 10^{11}$ N/m^2. The "tensile strength," defined as maximum stress before the breaking point, of the steel wire is about $(0.5 \sim 3) \times 10^9$ N/m^2. Since the stress is larger than the tensile strength, the wire will break. Typical tension on piano wires is about 500 N to 1000 N, and the wave speed on the string varies from 50 to 500 m/s.

C.   Speed of Sound Wave

A sound wave is a longitudinal pressure wave that travels in solid, liquid, or gas. It cannot travel in vacuum. The air at room temperature is approximately an ideal gas. Since air conducts heat poorly and the speed of sound wave is reasonably fast, the equation-of-state for sound wave propagation is $pV^\gamma = $ constant, where γ is the adiabatic gas constant, p is the air pressure, and V is the volume. The elastic modulus for air is $B \equiv -\frac{\Delta p}{\Delta V/V} = \gamma p$. The speed of sound wave becomes

$$v = \sqrt{\frac{B}{\rho}} = \sqrt{\frac{\gamma p}{\rho}} = \sqrt{\frac{\gamma n RT}{\rho V}} = \sqrt{\frac{\gamma RT}{M}}, \qquad (2.6)$$

where we apply the ideal gas law: $pV = nRT$. Here, $T = 273 + t_C$ is the absolute temperature, t_C is the temperature in degrees Celsius, $R = 8.314$ J/(mol · K) is the gas constant, n is the number of mole, and $M = \rho V/n$ the mass of the air per mol. Substituting the properties of

Air (21% oxygen, 78% nitrogen, $M = 0.02896$ kg/mol, $\gamma = 1.4$ for the diatomic molecules) in Eq. (2.6), we find

$$v = [331.3 + 0.6t_C] \text{ m/s}. \tag{2.7}$$

The sound speed in air depends essentially *only* on temperature. The sound propagates faster at higher temperature and propagates slower at lower temperature. The speed of sound in the ocean is a function of temperature, pressure (depth) and salinity as

$$v(t_C p(D), S) = 1449.2 + 4.6t_C + 0.055t_C^2 + 1.39(S-35) + 0.016D, \tag{2.8}$$

where v is sound speed in [m/s], t_C is water temperature in Celsius, D is depth below the surface in meters, and S is salinity in parts per thousand. Table 2.1 lists the sound speed in other media.

Table 2.1: Sound speed in m/s of some media at room temperature or temperature quoted.

Medium	Speed (m/s)	Sea water (10°C)	1,500	Diamond	12,000
Hydrogen (0°C)	1,286	Water (20°C)	1,500	Pyrex glass	5,600
Helium (0°C)	972	Mercury	1,450	Iron	4,510
Air (20°C)	343	Kerosene	1,300	Aluminum	5,100
Air (0°C)	331	Ethyl alcohol	1,200	Gold	2,030

★ Example 2.3: Calculate the wave speed and compare with properties of media listed below:

a) Helium: $\gamma = 5/3$, $M = 0.004$ kg/mole, $t_c = 20°C$;

Answer: $v = \sqrt{\frac{(1.666)(8.314)(293)}{0.004}} = 1000$ m/s

b) Water: $B = 2.2 \times 10^9$ Pa, $\rho = 1000$ kg/m³;

Answer: $v = \sqrt{\frac{2.2 \times 10^9}{1000}} = 1500$ m/s

c) Aluminum: $B = 7.0 \times 10^{10}$ Pa, $\rho = 2700$ kg/m³;

Answer: $v = \sqrt{\frac{7.0 \times 10^{10}}{2700}} = 5100$ m/s

II. Huygens' Principle and Properties of Wave Propagation in Media

To understand the mechanism of wave propagation, Christiaan Huygens surmised, "the wave fronts of a wave become the new sources of new disturbance as the wave travels in media," shown in Fig. 2.2(a). Each new sources radiate spherical waves outward. The superposition of these spherical wave fronts generates a new wave front of the new plane wave (dashed line). This is Huygens' Principle. This principle can explain all phenomena observed in wave propagation, including reflection, refraction, diffraction, and interference.

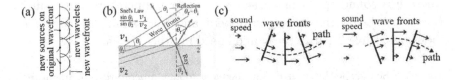

Figure 2.2: (a) Schematic drawing of wave propagation based on Huygens' Principle for plane waves in a straight line. (b) Refraction is the phenomenon in which a "wave changes direction in passing through two media at different wave speeds." Reflection is the phenomenon in which a "wave bounces back at the interface of two media having different wave impedance." Wave reflection obeys the law of reflection. (c) Refraction causes the wave to change its propagation direction in a medium in which the wave speed depends on location.

A. Wave Refraction

Refraction is the phenomenon in which a wave will bend toward the direction where the wave speed is slower. Figure 2.2(b) shows the refraction of a wave when passing through two media at different wave speeds, and Fig. 2.2(c) shows the wave bending towards the direction where the wave speed is slower. Typically, the air temperature near the ground is higher than that at higher altitude. The sound speed is thus higher near the ground. The sound wave will bend upward. However, when there is a temperature inversion, i.e. the temperature close to the ground is colder, and the temperature higher up is warmer, the sound speed at high altitude is faster than that near the ground, and the sound wave refracts or bends back towards the ground. Wind causes air molecules to move. Thus, the sound speed will increase in the direction of the wind and decrease opposite to the direction of the wind. If there is a wind gradient vs altitude, its effect resembles that of wave refraction. A sound wave will bend downward downwind and bend upward upwind.

B.   Wave Reflection

When a wave encounters the interface between two media with different acoustic impedance, wave reflection occurs. Figure 2.2(b) also shows wave reflection from the interface of two media. When the surface roughness is smaller than the wavelength, the wave reflection from the interface of two media obeys the *law of reflection*. This law states that

$$\theta_r = \theta_i \,, \tag{2.9}$$

where θ_r and θ_i are the reflection angle and incident angle, respectively. Wave reflection from a smooth surface is *specular* reflection; otherwise, it is a diffuse reflection, as shown in Fig. 2.3(A). Sound wave diffusers can minimize the effects of acoustic resonances inside a room.

Wave reflection can involve a phase change. Waves reflected from a fixed end have a 180° phase shift; however, waves reflected from an open end do not have a phase shift, as shown in Fig. 2.3(B). Sound wave reflection in a cylindrical pipe open to the air is a *pressure fixed end*, i.e. pressure cannot change at the position that is open to the air, and the pressure of the reflected wave changes sign (180° phase change). Reflection from a closed cylindrical pipe is a *pressure free end*, i.e. the pressure of the sound wave can increase to a maximum amplitude, and the pressure of the reflected wave does not change sign. A wave reflected from a wall is a reflection from a pressure free end, and there is no change of phase in the reflected pressure wave.

C.   Wave Diffraction

Wave diffraction is the phenomenon in which a wave can bend around an obstacle and spread outward. Waves at longer wavelengths will spread out more, while shorter-wavelength waves will spread out less. The

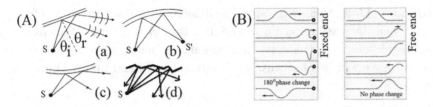

Figure 2.3: (A) Schematic drawing of wave reflection from different surfaces. Specular reflections are associated with smooth surfaces such as (a)–(c), while diffuse reflection is associated with a rough surface, such as (d). All reflections obey the law of reflection in Eq. (2.9). (B) Sequences of wave pulses vs time for wave reflection from a fixed-end and free-end boundary conditions. The reflected wave change phase by 180° from a fixed end, while wave reflection from a free end does not change the phase.

wave's angular spread θ, due to diffraction, is

$$a \sin \theta \sim \lambda = v/f, \qquad (2.10)$$

where a is the diameter of the obstacle or size of a hole opening, θ is the angular spread of the diffracted wave, λ is the wavelength, and v is the speed of sound. If $\lambda \ll a$, the angular spread is small and there is little diffraction. If $\lambda \geq a$ or $f \leq v/a$, the wave is prone to diffraction phenomenon as shown schematically in Fig. 2.4. Our ears can differentiate the direction of high-frequency sounds coming from in front of or behind us more easily than low-frequency sounds because high-frequency sounds coming from behind us can cast a shadow via the ear pinna.

Figure 2.4: Schematic drawing of a plane wave passing through an obstruction (pinhole or a barrier), displaying the diffraction phenomenon. Diffraction is important when the wavelength λ is larger than the aperture a.

The diffraction of a *pure tone* with wavelength comparable with the slit width or the diameter of a circular aperture can produce constructive and destructive interference of the diffracted wave. Figure 2.5(a) shows the diffraction pattern of a wave passing through a circular

Figure 2.5: (a) Schematic drawing of the intensity distribution due to diffraction through a slit or a pin pole. (b) Combination of wave diffraction and wave interference can produce the double-slit interference pattern.

aperture. When a pure tone wave passes through two slits with slit spacing comparable to the wavelength, interference can produce a constructive and destructive interference pattern, shown schematically at Fig. 2.5(b) (see also Example 2.5).

D.   Wave Superposition, Interference

When waves overlap, the disturbances to the medium can add or subtract, i.e. they interfere. Wave interference can be constructive or destructive. Constructive interference leads to a larger wave amplitude, and destructive interference results in a smaller wave amplitude. Figure 2.5(b) shows the interference pattern of waves from a double-slit experiment. Constructive interference occurs at the peak intensity locations, while destructive interference occurs at the minimum intensity locations. Figure 2.6(a) illustrates the constructive and destructive interference of wave pulses. When wave pulses overlap, they can interfere constructively to give a larger pulse or interfere destructively to give zero disturbance at the instant and location of wave overlap.

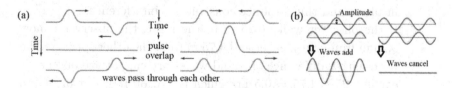

Figure 2.6: (a) Schematic drawing of two wave pulses passing through each other vs time. When these two waves overlap, they can interfere destructively or constructively. Two pulses pass through each other without changing their shapes. (b) Sine waves at the same frequency interfere constructively for 0° phase difference, or destructively at 180° phase difference. Partial interference occurs when their phase difference is neither 0° nor 180°.

Figure 2.6(b) shows interference of two sinusoidal waves. When two sinusoidal waves are in phase, they will interfere constructively to give twice the wave amplitude or 4 times the intensity. This occurs when the path length of two waves differs by 0, $\pm\lambda$, $\pm2\lambda$, ... When two sinusoidal waves are 180° out of phase, they interfere destructively and cancel each other [see the right plot in Fig. 2.6(b)]. This happens when two waves have $\pm\frac{1}{2}\lambda$, $\pm\frac{3}{2}\lambda$, $\pm\frac{5}{2}\lambda$, ..., in path length difference.

★ Example 2.4: Two identical loudspeakers are facing each other at a distance of 10.0 m, shown below. They emit the same pure tones "in phase" at $f = 343$ Hz. The air temperature is 20°C. (1) What is the wavelength of the sound wave? (2) Will an observer positioned in the middle between the two speakers, i.e. 5 m from each speaker, hear constructive or destructive interference? (3) What happens if the observer moves 0.25 m toward one of the speakers? (4) What happens if the observer moves 0.50 m toward one of the speakers?

Answer:

(1) At 20°C, the sound wave speed is $v = 343$ m/s. Since the loud speakers are emitting a pure tone of 343 Hz, the wavelength of the sound wave in air is $\lambda = v/f = 1$ m.

(2) At the middle, the sound waves from both speakers will reach the observer with zero path-length difference, meaning they are in phase and thus produce constructive interference.

(3) If the observer walks 0.25 m to the right, the observer is now 5.25 m from one speaker and 4.75 m from the other speaker. The path-length difference of sound waves arriving at the observer's ear is $5.25 - 4.75 = 0.5$ m, which is $\frac{1}{2}\lambda$, or 180° out of phase. The interference is destructive.

(4) If, instead, the observer walks 0.50 m to the right, the observer's distances from the speakers are 5.5 m and 4.5 m, respectively. The path difference of the two sources is $5.5 - 4.5 = 1.0$ m, which is equal to λ. The interference is constructive.

In general, if the person walk Δx distance from 1 speaker, the path-length difference from two coherent sources is $(5m + \Delta x) - (5m - \Delta x) = 2\Delta x$. For destructive interference, the path-length difference will be $2\Delta x = \frac{1}{2}\lambda, \frac{3}{2}\lambda, \ldots$. For constructive interference, the path-length difference is $2\Delta x = 0, \lambda, 2\lambda, 3\lambda, \ldots$. This agrees with the analysis in items (2) and (3) above.

★ Example 2.5: Consider two identical loudspeakers emitting a pure tone at $f = 343$ Hz. The air temperature is 20°C. An observer at position C_0 in the middle between the two speakers is located at a distance of 8.0 m, as indicated below, will hear constructive interference. A series of positions at $D_{\pm 1}$, $C_{\pm 1}$, $D_{\pm 2}$, ... will observe interference that is destructive, constructive, destructive, etc., resembling the pattern in Fig. 2.5(b). What are the distances y from C_0 for the $D_{\pm 1}$, $C_{\pm 1}$ positions?

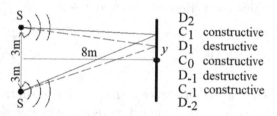

Answer: First, the wavelength is $\lambda = \frac{v}{f} = 1$ m, as discussed in Example 2.4. Figure 2.5(b) shows this series of constructive and destructive interference pattern, $C_0, D_{\pm 1}, D_{\pm 1}, \ldots$.

(1) Destructive D_1 and D_{-1} positions: The path difference from the two speakers must be $\frac{1}{2}\lambda$ for destructive interference, i.e. $\sqrt{(3 + y_{D1})^2 + 8^2} - \sqrt{(3 - y_{D1})^2 + 8^2} = \pm\frac{1}{2}\lambda$. Here we use the Pythagorean Theorem for the length of the hypotenuse. Solving this algebraic equation, we find $y_{D\pm 1} = \pm 0.714$ m. These destructive positions are symmetric with respect to the center C_0 position.

(2) Constructive C_1 and C_{-1} positions: The path difference from the two speakers must be λ for constructive interference, i.e. $\sqrt{(3 + y_{C1})^2 + 8^2} - \sqrt{(3 - y_{C1})^2 + 8^2} = \pm\lambda$. Solving this algebraic equation, we find $y_{C\pm 1} = \pm 1.442$ m.

★ Example 2.6: Waves (a), (b), and (c) are wave disturbance vs time at a location. Find the superposition of waves (a) + (b); and waves (a) + (c).

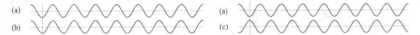

Answer: The waves (a) and (b) are in-phase and the addition of these waves will be constructive. On the contrary, the waves (a) and (c) are out-of-phase, or phase shifted by 180°. The superposition of waves (a) and (c) will produce destructive interference and cancel each other.

★ Example 2.7: Waves A and B in the following drawing are wave disturbance vs time at a location. The period of wave A is 4 unit of time, and B is 2 unit of time. Use a pencil to add waves A and B, and determine the period of the wave A + B.

Answer: The wave interference is the addition or subtraction of oscillation amplitudes. The summed wave is arithmetic addition of the displacements of these two waves, shown in the right plot. Since the period of wave A is four units of time, and wave B has a period of two units, the period of the combined wave is four units of time. The period of the summed wave is the smallest number that is divisible by both periods. In this example, 4 is divisible by both 4 and 2. Thus the summed wave has a period of 4 units of time.

E.   Beat

The linear superposition of two waves at frequencies f_1 and f_2, with a small frequency difference between them, produces a resultant wave that has a "fusion" frequency and a "beat" frequency. The fusion frequency is the average of the two frequencies, and the beat frequency is the difference of the two frequencies, i.e.

$$f_{\text{fusion}} = \frac{1}{2}(f_1 + f_2), \quad f_{\text{beat}} = |\Delta f| = |f_1 - f_2|. \qquad (2.11)$$

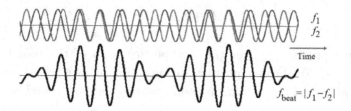

Figure 2.7: Top: schematic drawing of the combined tone of two pure tones at frequencies f_1 and f_2 and with equal amplitude. Bottom: the combined wave has a fusion frequency $f_{\text{fusion}} = \frac{1}{2}(f_2 + f_1)$ with amplitude varying at the beat frequency of $f_{\text{beat}} = |f_1 - f_2|$.

The "beat" is the interference of waves in "time" with the amplitude modulation of the summed wave, as shown in Fig. 2.7. Typically, when the beat frequency is less than 10 Hz, we can hear the beat, and when the beat frequency is greater than 15 Hz, the beat becomes "rough" to our ears. The roughness can produce dissonance of two pure tones.

★ Example 2.8: When you hit the piano middle-A key and a 440-Hz tuning fork simultaneously, you hear a beat at 2 Hz. What are possible frequencies of the piano key?

  Answer: Since $f_{\text{beat}} = |440 - f| = 2$ Hz, possible frequencies are either $f = 438$ Hz or 442 Hz.

★ Example 2.9: What will we hear when playing two pure tones at frequencies 398 and 402 Hz?

  Answer: We hear a tone at fusion frequency $\frac{1}{2}(f_2 + f_1) = 400$ Hz, and a beat frequency 4 Hz.

F. Intensity of Combined Incoherent Sounds

"Coherent" sound sources have correlated phases. The interference of coherent sources is the addition or subtraction of oscillation displacements. On the other hand, the phases of "uncorrelated and incoherent" sound sources do not have a fixed correlation. The interference of uncorrelated and incoherent sound sources produces a summed intensity that is the sum of the intensities of the individual sound sources:

$$I = I_1 + I_2 + I_3 + \cdots = \sum_i I_i . \qquad (2.12)$$

Let Δp represent the rms pressure perturbation of the sound wave. From Eq. (2.5), the rms pressure perturbation of each sound source is $\Delta p_i^2 = \rho v I_i$, and the net rms pressure of the combined sound is

$$\Delta p^2 = \Delta p_1^2 + \Delta p_2^2 + \Delta p_3^2 + \cdots = \sum_i {\Delta p_i}^2 . \qquad (2.13)$$

The combined rms pressure amplitude is

$$\Delta p = \sqrt{\Delta p_1^2 + \Delta p_2^2 + \Delta p_3^2 + \cdots} .$$

III. Doppler Effect

The Doppler Effect is the "apparent" frequency shift observed when a sound source and an observer are in relative motion, theorized by Christian Doppler in 1842. In 1845, Buys Ballot experimentally verified the Doppler Effect. When the sound source and an observer are moving towards each other, the frequency appears higher; when the sound source and the observer are moving away from each other, the frequency appears lower. Figure 2.8(a) schematically shows the wave propagation of a stationary source, where each circle represents the wave front of one wavelength.

Figure 2.8(b) shows an observer O moving toward a stationary source. The observer will encounter the wave fronts at a smaller apparent period T'. Figure 2.8(c) shows the source moving toward a

Figure 2.8: (a) Schematic drawing of a stationary source emitting waves outward. The wavelength is equal to the distance between wave crests, i.e. $\lambda = vT$, where v is the speed of the wave and T is the wave period of the stationary source. (b) Schematic drawing of an observer (O) moving toward the source. (c) Schematic drawing of a source moving toward a stationary observer. (d) Schematic drawings of shock wave formation when the speed of the source is equal to or greater than the speed of the wave.

stationary observer. The apparent wavelength λ' becomes smaller in the direction of motion, thus the frequency appears higher. When the speed of the source is faster than the speed of wave, the wave fronts appear clump together, and a shock wave occurs as shown in Fig. 2.8(d).

A. Observer in Motion

When the observer is moving toward a stationary source, the observer will encounter wave fronts at a shorter apparent wave period, and the apparent frequency will be higher. The time (or the perceived period) that the observer encounters wave crests is $T' = \frac{vT}{v+v_o}$. The frequency the observer perceives is $f' = \frac{1}{T'} = f_s \frac{v+v_o}{v}$, where $f_s = \frac{1}{T}$ is the source frequency, v is the wave speed, v_o is the velocity of the observer, and f' is the apparent frequency. On the other hand, if the observer is moving away from the source, the apparent period lengthens, and the apparent frequency shortens. The perceived period and the apparent frequency are $T' = \frac{vT}{v-v_o}$ and $f' = \frac{1}{T'} = f_s \frac{v-v_o}{v}$. Combining of both cases, we find $f' = f_s \frac{v\pm v_o}{v}$, where the $\pm$ sign corresponds to an observer moving toward/away from the source.

B. Source in Motion

Figure 2.8(c) shows the wave fronts of a moving source. For the observer in the direction of the moving source, the successive wave fronts appear squeezed, and wavelength appears shorter. The apparent frequency increases. The apparent wavelength is $\lambda' = (v - v_s)T$, and the perceived frequency f' becomes $f' = \frac{v}{\lambda'} = f_s \frac{v}{v-v_s}$, where v is the wave speed and v_s is the velocity of the source. When the source is moving away from the observer, the apparent wavelength is stretched to $\lambda' = (v + v_s)T$. The perceived frequency then becomes $f' = \frac{v}{\lambda'} = f_s \frac{v}{v+v_s}$. Whence, the Doppler effect for the source in motion is $f' = f_s \frac{v}{v\mp v_s}$, where the $\mp$ sign corresponds to the source moving toward/away from the observer.

The resulting Doppler Effect for the observer and the source in relative motion is

$$f' = \frac{v \pm v_o}{v \mp v_s} \times f_s \qquad (2.14)$$

where v is the speed of sound in the medium, $\pm v_o$ in the numerator corresponds to the observer moving toward or away from the source; and $\mp v_s$ in the denominator corresponds to the source moving toward or away from the observer.

★ Example 2.10: A monorail approaches a platform at a speed of 10 m/s while it blows its whistle. A musician with perfect pitch hears the whistle as "middle C" at 261 Hz. There is no wind, and it is 0°C. What is the frequency of the whistle when the monorail is at rest?

   Answer: At 0°C without wind, the speed of sound is $v = 331$ m/s. The train whistle is moving at $v_s = 10$ m/s toward the observer. Using Eq. (2.14), we find $261 = \frac{331}{331-10}f_s$, or $f_s = 253$ Hz.

C.   Shock Wave

When the speed of a sound source is faster than the wave speed of the transmission medium, a cone-shaped shock wave appears. The cone-shaped shock wave creates a sonic boom as it passes stationary observers. Figure 2.8(d) shows the shock wave front. The shock wave subtends an angle θ at $\sin\theta = v/v_s \equiv 1/(\text{Mach number})$, where the Mach number is the ratio of source speed to the wave speed. Shock waves can cause damage when a meteorite strikes the Earth at high speed. A meteorite entered the Earth's atmosphere over Chelyabinsk, Russia on Feb. 15, 2013 at an altitude of 29.7 km and a speed of 19.2 km/s. An estimation put the meteor at 20 m in diameter and 12,000 tons. Its shock wave shattered windows within a 55-mile radius of its path.

IV.   Resonances and Standing Wave Modes

Resonances exist in all physical systems, such as musical instruments, buildings, bridges, etc. The wave pattern of a resonance is the standing wave. Resonance had caused the collapse of the old Tacoma Narrows Bridge in 64 km/h winds on the morning of Nov. 7, 1940. Resonance arises from constructive interference between the waves that travel in the system and their reflections from the boundaries of the system. At these resonant frequencies, the wave amplitudes build up and become stationary. At non-resonant frequencies, the interference is non-constructive and the wave dissipates. Musical instruments rely on resonances to produce their quality of sound; human speech relies on resonances to produce differentiable vowel sounds.

A.   Standing Waves on Strings

Since the string of string instruments are fixed at both ends (the nut and the bridge support), a wave reflecting from the fixed end changes phase by 180°. When the incoming and reflected waves overlap, interference may be constructive (leading to larger wave amplitude) or destructive (leading to smaller wave amplitude). This interference produces "standing wave modes" or "resonances" on the string, shown schematically in Fig. 2.9 for mode number $n = 1, 2, 3, 4$ at an instant of time. The string can be at a different line at a different instant of time. The time for completing one cycle is the period T, and the frequency is the number of cycles per second, or $f = 1/T$.

$$
\begin{array}{cccc}
n = \ 1 & 2 & 3 & 4 \\
\lambda = \ 2L & L & 2L/3 & 2L/4
\end{array}
$$

Figure 2.9: Schematic drawing of the standing wave at an instant of time for a string with a boundary condition fixed at both ends. The string does not move at "nodal positions," and the string vibrates at a maximum amplitude at "antinodes."

A standing wave is the stationary wave-like pattern that results from the interference of two or more traveling waves. Standing waves are resonances of a string vibrating system having regions of minimum amplitude called nodes and regions of maximum amplitude called antinodes. The wavelength of the lowest standing wave mode is $\lambda_1 = 2L$. The wavelength of the 2nd mode is $\lambda_2 = \frac{2L}{2} = L$, and the wavelength of the 3rd mode is $\lambda_3 = \frac{2L}{3}, \ldots$, etc. The standing wave frequencies on a string of length L become

$$
f_n = \frac{v}{\lambda_n} = n\frac{v}{2L}, \quad n = 1, 2, 3, 4, \ldots, \tag{2.15}
$$

where $v = \sqrt{F_T/\mu}$ is the transverse wave speed on the string [Eq. (2.3)], F_T is the tension on the string, $\mu \equiv M/L$ is the mass per unit length

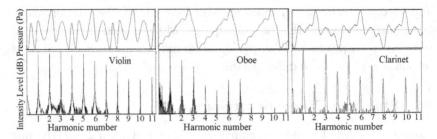

Figure 2.10: (Top) Examples of waveforms produced by a violin, oboe, and clarinet. (Bottom) The corresponding frequency spectra. The waveforms are periodic, and the spectra contain a fundamental mode and all integer harmonics.

with M the mass and L the length, and n is the mode number. The spectrum of musical instruments will contain all harmonics that the instrument can sustain. Figure 2.10 shows waveforms and spectra of music notes from a violin, oboe and clarinet.

★ Example 2.11: A 60 cm long nylon guitar string has mass of 6.0×10^{-4} kg. What are the frequencies and wavelengths of the 3 lowest resonance modes under a tension of 70 N?

Answer: The mass per unit length of the nylon string is $\mu = \frac{0.0006\,\text{kg}}{0.6\,\text{m}} = 0.001\frac{\text{kg}}{\text{m}}$. Thus, the speed of the wave on the string is $v = \sqrt{70\,\text{N}/(0.001\,\text{kg/m})} = 265$ m/s. The wavelength of the fundamental mode is $\lambda_1 = 2L = 1.20$ m. The fundamental frequency is $f_1 = \frac{v}{\lambda_1} = 220$ Hz. The frequencies and wavelengths of the 2nd and 3rd harmonics are $f_2 = 2f_1 = 440$ Hz, $\lambda_2 = 0.6$ m; and $f_3 = 3f_1 = 660$ Hz, $\lambda_3 = 0.4$ m respectively. Note that the wave speed is $v = f_1\lambda_1 = f_2\lambda_2 = f_3\lambda_3$.

★ Example 2.12: Find the first 3 modes of vibration of a steel guitar string with diameter 0.30 mm, length 65 cm and under a tension of $F_T = 100$ N. (Density of steel is $\rho = 7700$ kg/m^3)

Answer: The mass of a steel string of length L is $M = \rho V = \rho \pi r^2 L$. The mass per unit length is $\mu = M/L = \rho \pi r^2 = 5.4 \times 10^{-4}$ kg/m. The wave speed on the string is $v = \sqrt{100/(5.4 \times 10^{-4})} = 429$ m/s. The wavelength of the fundamental mode is $\lambda_1 = 2L = 1.3$ m. The frequency of the fundamental mode is $f_1 = v/\lambda_1 = 429/1.3 = 330$ Hz. The frequencies of the second and third harmonics are $2 \times 330 = 660$ Hz and $3 \times 330 = 990$ Hz, respectively.

★ Example 2.13: A 60-cm long nylon guitar string has a mass of 6.0×10^{-4} kg, and the tension on the string is 70 N. The guitar is producing its fundamental mode frequency of 220 Hz. What is the wavelength of the sound wave it is producing in a room at 20°C?

Answer: The speed of sound wave in the room at 20°C is 343 m/s. The wavelength of the sound wave in air is $\lambda = 343/220 = 1.56$ m. On the other hand, the wavelength of the transverse wave on string at $\lambda_1 = 2L = 1.20$ m. The frequency of the string excites the guitar sound box at the frequency of string vibration, and in turn, excite the air molecules at the same frequency. The wavelengths of these two media are different, but the frequency in these two media is identical.

B. Standing Wave Inside Cylindrical Air Columns

Sound wave is a pressure wave due to oscillations of air molecules. The pressure in the sound wave is 90° out of phase with respect to the displacement of air-molecules [see Eq. (2.4) and shown schematically in Fig. 2.11(a)]. The *closed-end of an air column* is *pressure antinode*; and the *open-end of air-column* is a *pressure node (nodal point)*. Figures 2.11(b) and (c) shows possible standing wave modes of the open-open and open-closed pipes. Possible wavelengths and frequencies of standing wave modes are

$$\text{Open: } \lambda_n = \frac{2L}{n}; \quad f_n = \frac{v}{\lambda_n} = n\frac{v}{2L}, \quad n = 1, 2, 3, 4, 5, 6, \ldots \text{ (2.16)}$$

$$\text{Closed: } \lambda_n = \frac{4L}{n}; \quad f_n = \frac{v}{\lambda_n} = n\frac{v}{4L}, \quad n = 1, 3, 5, 7, \ldots \quad \text{(2.17)}$$

The speed of sound wave in air is $v = [331.3 + 0.6t_C] \frac{m}{s}$, where t_C is the temperature in Celsius [see Eq. (2.7)].

★ Example 2.14: What is the property of the sound wave pressure at the open/close end of a cylindrical pipe?

Answer: At the open end of the pipe, it is a fixed point of the pressure wave. At the closed end of the pipe, the wave pressure can reach a maximum amplitude. It is a free end boundary condition of the pressure wave [see the location A in Fig. 2.11(a)].

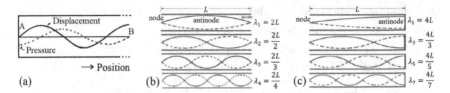

Figure 2.11: (a) The pressure and displacement vector of air molecules are 90° out of phase. Position A is a pressure antinode, and position B is a pressure node. (b) Possible standing pressure waves for an open pipe where air column is open on both ends. (c) Standing wave modes of a closed pipe where air column is closed on one end.

★ Example 2.15: Find the first three resonances for an open pipe 0.75 m long at 20°C.

Answer: The speed of sound in air at 20°C is 343 m/s. The wavelength of the fundamental mode of the open pipe is $\lambda_1 = 2L = 1.5$ m. The fundamental mode frequency is $f_1 = v/\lambda_1 = 228$ Hz. The resonant frequencies of the 2nd and 3rd harmonics are $2f_1 = 556$ Hz and $3f_1 = 684$ Hz.

★ Example 2.16: Find the first three resonances for a closed cylindrical pipe 0.025 m long at 30°C.

Answer: The speed of sound in air at 30°C is 350 m/s. The wavelength of the fundamental mode of the closed pipe is $\lambda_1 = 4L = 0.10$ m. The fundamental mode frequency is $f_1 = \frac{v}{\lambda_1} = 3500$ Hz. The frequencies of the next 2 harmonics are $3f_1 = 10{,}500$ Hz and $5f_1 = 17{,}500$ Hz.

The open tube has all integer harmonics. The spectrum of oboe shown in the middle plot of Fig. 2.10 has all harmonics, and thus the oboe is an open pipe. The closed pipe has only odd harmonics f_1, f_3, $f_5 \ldots$ i.e. there is no even harmonics. The spectrum of a clarinet shown in the right plot of Fig. 2.10 has mainly odd harmonics. The amplitudes of even harmonics 2, 4, and 6 of the clarinet are much lower than that of odd harmonics by more than 20 dB, i.e. their intensities are only $\frac{1}{100}$ of those of the odd harmonics. The clarinet is a closed cylindrical pipe. However, we note that higher-order even harmonics do exist in the clarinet spectrum.

C.   Resonances in Singing Rod, Membranes, Helmholtz Resonators, and Electronic Circuits

The speed of a sound (longitudinal) wave in a solid is $v = \sqrt{B/\rho}$, where B is the elastic modulus and ρ is the density. The wavelength of the fundamental mode is $2L$, where L is the length of the rod. The frequencies of the longitudinal standing wave are $f_n = n\frac{v}{2L}$. Figure 1.10 shows some transverse standing wave modes of a thin rod at a fixed-end or free-end boundary conditions.

Two-dimensional (2D) standing waves occurs also in extended membranes such as drum and music resonance boxes. Chladni wave patterns are resonances of 2D systems. They can enhance the sound qualities of musical instruments.

A cavity that fills with air is the Helmholtz resonator, shown in Fig. 1.6 with resonant frequency given by Eq. (1.17). Before the invention of microphones, amplifiers, and spectrum analyzers, scientists used Helmholtz resonators to study resonances of vibrational objects and analyze complex sounds. The mufflers of mechanical systems use Helmholtz resonators to absorb noise (unwanted) sounds at their resonance frequencies.

Electronic circuits with resistors, inductors and capacitors, called RLC circuits, also exhibit resonance phenomenon. The resonance frequency of an RLC circuit is $f_0 = \frac{1}{2\pi\sqrt{LC}}$, where L is the inductance, and C is the capacitance. The symbol R represents the resistance for power load and dissipative elements. Chapter 8 discusses electronic circuits in acoustics.

D.   Resonance and Ultrasound Generation

In 1882, Jacques and Pierre Curie discovered the piezoelectric phenomenon in quartz, topaz, schorl tourmaline, cane sugar, Rochelle salt, etc. When a slice of piezoelectric crystal is compressed, it produces electricity along a certain axis. The reversible process of producing sound waves by applying an electric field to the piezoelectric crystals also works. The sound production is enhanced by the resonance condition, e.g. $f = \frac{v}{2l}$, where v is the speed of a sound wave in the crystal and l is the length of the crystal. For quartz, the speed of a sound wave is 5.4×10^3 m/s along the perpendicular axis, and 6.0×10^3 m/s along the parallel axis. Another physics phenomenon useful for the ultrasound transducer is "magnetostriction," which is where ferromagnetic materials can change length along the direction of the applied magnetic field. Resonance can also enhance ultrasound production.

E.   Characteristics of Resonances and the Q-factor

When the frequency of the driving force is the same as the "natural" or resonant frequency of an oscillating system, the amplitude of the oscillation reaches its maximum. Figure 2.12 shows the universal response of all resonances, characterized by a resonance frequency f_0, a bandwidth Δf, (defined as $\Delta f = f_2 - f_1$ of frequencies at $\frac{1}{\sqrt{2}} = 71\%$ of the maximum amplitude). As discussed in Appendix B(3), the Q-factor or quality-factor of the resonance is $Q = f_0/\Delta f$.

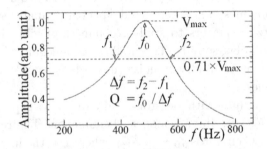

Figure 2.12: Relative response amplitude of resonance vs driven frequency. The frequency at the maximum of the response curve is the resonance frequency f_0. The frequencies that the response is 71% of the maximum amplitude are cutoff frequencies: f_1 and f_2. The bandwidth of the resonance is $\Delta f = f_2 - f_1$, and the quality-factor of the resonance is $Q = f_0/\Delta f$.

V.   Sound Intensity and Acoustic Impedance

Waves transport wave energy and wave momentum along the medium. A sound wave arises from oscillations of air molecules as the wave propagates in the medium. The oscillation of air molecules in a sound wave's propagation is similar to that of simple harmonic oscillators. The total energy density of these molecules is the sum of kinetic energy and potential energy densities [Eq. (2.4)]. The energy is transported as the wave energy. The energy density (energy per unit volume, or [J/m³]) and the sound intensity I (power per unit area, or $[\text{W/m}^2]$) are respectively

$$w = \frac{1}{2}\rho u^2 + \frac{1}{2\rho v^2}\Delta p^2 = \frac{1}{2}\rho u_{\text{max}}^2 = \frac{1}{2\rho v^2}\Delta p_{\text{max}}^2 \qquad (2.18)$$

$$I = wv = \frac{1}{2\rho v}\Delta p_{\text{max}}^2 = \frac{1}{\rho v}\Delta p_{\text{rms}}^2 = \frac{1}{2}\rho v u_{\text{max}}^2 = 2\pi^2 \rho v f^2 x_0^2 \quad (2.19)$$

where u is the velocity of molecules, Δp is the pressure of the sound wave in the medium, ρ is the density of the medium, v is the speed of the wave, and Δp_{max} is the pressure amplitude of the wave. The maximum speed of the molecule in the medium is $u_{\text{max}} = 2\pi f x_0$, where f is the frequency and x_0 is the amplitude of molecular motion in the sound wave. Since $\Delta p_{\text{max}} = \rho v u_{\text{max}}$, we find

$$\Delta p_{\text{max}} = \sqrt{2}\Delta p_{\text{rms}} = 2\pi f \rho v x_0 . \qquad (2.20)$$

The quantity ρv is the "intrinsic acoustic impedance," analogous to impedance in electric circuitry. Using air density $\rho = 1.204 \text{ kg/m}^3$ and $v = 343$ m/s at 20°C, we find $\rho v = 413 \text{ Ns/m}^3$, commonly *set* at 400 Ns/m^3 for convenience.

★ Example 2.17: The threshold intensity of human hearing is 10^{-12} W/m^2 at 1000 Hz. What is the maximum speed and displacement of air molecules for this sound wave?

Answer: Using Eq. (2.19) and $\rho v = 400$ kg/m^{2}s for air, we find $10^{-12} = 2\pi^2(400)(1000)^2 x_0^2$, or $x_0 = 1.1 \times 10^{-11}$ m that is less than the size of a molecule ($\sim 10^{-10}$ m). The maximum speed of molecules in the sound wave propagation is $u_{\text{max}} = 2\pi f x_0 = 7.1 \times 10^{-8}$ m/s.

★ Example 2.18: The threshold of human hearing is $p_{\text{rms}} = 20$ μPa at $f = 3000$ Hz. What is the displacement amplitude of air molecules at this pressure?

Answer: Using Eq. (2.20), we find $\sqrt{2}\times 20$ μPa $= 2\pi \times 3000 \times 400 \times x_0$, so $x_0 \sim 3.75 \times 10^{-12}$ m. Note that the size of atoms is typically 10^{-10} m, and the average distance between adjacent air molecules is about 35×10^{-10} m. Our ears are indeed very sensitive! At 3000 Hz, the displacement amplitude is 3 times smaller than that in Example 2.17.

★ Example 2.19: What is the displacement amplitude of air molecules at the threshold of pain occurring at sound intensity $I = 10$ W/m^2, or $p_{\rm rms} = 63.2$ Pa at $f = 1000$ Hz?

Answer: At the threshold of pain, the acoustic wave pressure is $\Delta p_{\rm rms} = 63.2$ Pa $= \sqrt{2}\,\pi f \rho v x_0$ and the maximum displacement of air molecules is $x_0 = 36\ \mu$m. For comparison, the average diameter of a thread of our hair is 50–100 μm. At sound intensity $I = 10^4$ W/m^2, or $\Delta p_{\rm rms} = 2000$ Pa, our eardrum may rupture instantly, and the amplitude of air-molecule vibration is 1.1 mm.

★ Example 2.20: At the threshold of pain, the intensity of sound is about 1.0 W/m^2. If this sound energy is perfectly absorbed in a cup with area 50 cm^2, how long would it take to heat up 100 cm^3 of water from 20°C to 70°C? [The specific heat of water is 4.186 J/(g °C).]

Answer: The density of water is 1 g/cm^3. The mass of 100 cm^3 water is Mass = (density) × (volume) = (1 g/cm^3) × (100 cm^3) = 100 g. The amount of heat needed is (mass) × (specific heat) × (difference in temperature) = (100 g) × (4.186 J/g°C) × (70°C − 20°C) = 2.093 × 10^4 J. The total sound energy is (Intensity) × (Area) × (time), or 2.093 × 10^4 J = (1.0 W/m^2) × (50 × 10^{-4} m^2) × (time). Thus, we find the "time" = 4.186 × 10^7 s, which is about 485 days. The acoustic energy is small.

A.   Acoustic Impedance

The acoustic impedance is the ratio of sound pressure to the volume flow rate U,

$$Z_A \equiv \frac{\Delta p}{U} = \frac{\rho v}{A}, \qquad (2.21)$$

where $\Delta p = \rho v u$ is the wave pressure, $U = Au$ is the volume flow rate of the medium, ρ is the density, v is the speed of the sound wave, u is the speed of molecules, and A is the cross-sectional area. When sound waves pass through media with different acoustic impedance, wave reflection occurs at the boundary of the two media. When an acoustic wave reaches the end of a pipe of a musical instrument, the impedance in the pipe differs from that of outside due to the difference in cross-sectional areas. Thus, the wave will reflect back into the pipe. The amplitude of the reflected wave depends on the difference in impedance between inside and outside the pipe. A pipe with a smaller cross-sectional area will reflect more from the end. Interference between the outgoing wave and the reflected wave inside the pipe produces standing waves, discussed in Sec. IV.

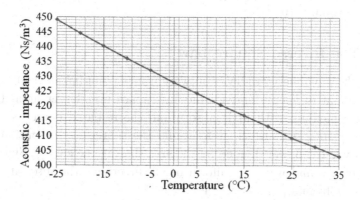

Figure 2.13: The acoustic impedance of air at various temperature. For convenience, it is normally set at $\rho v \equiv 400$ Ns/m^3.

The *intrinsic* acoustic impedance of a plane wave is $Z \equiv Z_A A = \frac{\Delta p}{u} = \rho v$. Using $\rho = 1.29$ kg/m^3 for the density of air and $v = 343$ m/s for the speed of sound at 20°C and one atmospheric pressure, the intrinsic acoustic impedance is $Z = 413$ Ns/m^3, normally set at $Z = 400$ Ns/m^3. Figure 2.13 shows the intrinsic acoustic impedance of air vs temperature. The (average) sound intensity and the rms pressure are

$$I = \frac{\Delta p_{\text{rms}}^2}{Z} = \frac{\Delta p_{\text{max}}^2}{2Z} \quad \text{and} \quad \Delta p_{\text{rms}} = \sqrt{IZ}. \tag{2.22}$$

★ Example 2.21: What is the rms pressure of a sound wave at $I_0 = 10^{-12}$ W/m^2 (threshold of hearing) and 1.0 W/m^2 (threshold of pain)?

Answer: Using Eq. (2.22), we find $\frac{1}{400}\Delta p_{\text{rms}}^2 = 10^{-12}$ W/m^2 at the threshold of hearing, whence $\Delta p_{\text{rms}} = 20$ μPa. This pressure is set as the reference pressure for the definition of the sound pressure level in Chapter 3. Similarly, $\frac{1}{400}\Delta p_{\text{rms}}^2 = 1.0$ W/m^2 at the threshold of pain, and we obtain $\Delta p_{\text{rms}} = 20$ Pa.

★ Example 2.22: The area of an average human eardrum is 5.0×10^{-5} m^2. What is the force on the eardrum if the sound pressure is 60 μPa and 60 Pa?

Answer: Force = (pressure) × (Area) = $(60\ \mu\text{Pa}) \times (5.0 \times 10^{-5}\ \text{m}^2) = 3.0 \times 10^{-9}$ N. If the pressure is increased to 60 Pa, the amount of force is 3.0×10^{-3} N.

B.   The Inverse Square Law

When the wave energy radiates outward from a point (small size) source, this amount of energy spreads out into space ever larger area. In an isotropic medium with little absorption, the radiated energy spreads uniformly into the surface area around the source. By definition, the intensity is the power divided by the area sharing the energy. Since the speed of a sound wave is uniform in all directions, the intensity at a distance r from a point source is

$$I \equiv \frac{\text{Power}}{\text{Area}} = Y\frac{W}{4\pi r^2}, \tag{2.23}$$

where W is the power of the source, Y is the directivity parameter of the source, and $4\pi r^2$ is the area of a spherical shell of radius r. This is the *inverse square law*. Intensity is reduced by a factor of 4 by doubling the distance from the source. Figure 2.14 shows the inverse square law for various directional geometries.

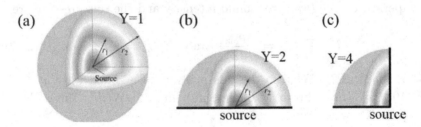

Figure 2.14: Schematic drawing of a point source emitting waves spreading outward through a space into the area of a sphere $4\pi r^2$. (a) The power spreads uniformly into space. (b) The source is located on a non-absorbing floor, where the energy spreads into a hemisphere. (c) The source is located on the side of a non-absorbing wall and on a non-absorbing floor. The energy spreads outward into $\frac{1}{4}$ of the spherical space, and thus the directivity factor is 4.

If a point source is located outdoors up on a platform, the directivity parameter is $Y = 1$, i.e. the sound energy spreads uniformly outwards over the spherical shell. If the source is located at an open non-absorbing concrete floor with little acoustic absorption, the sound power spreads outward only in a hemisphere, and the directivity parameter is $Y = 2$. If the sound source is located at the floor next to a wall, the directivity is $Y = 4$; and if the sound source is located at the corner of a room bounded by 3 walls of high reflectivity, the directivity parameter is $Y = 8$. The directivity parameter $Y > 8$ is possible for some directional sources, e.g. directional speakers. The sound waves from most musical instruments, including the trumpet, spread out in all directions, and the sound intensity obeys the inverse square law with $Y = 1$.

★ Example 2.23: The refractive index of light depends on the air density and temperature, given by $n(P,T) = 1 + 0.000293 \frac{p}{p_0} \frac{T_0}{T}$, where p and T are the pressure and temperature on the absolute scale, respectively, and $p_0 = 1$ atm and $T_0 = 273$ K. Photographic film can record the propagation of sound waves. The photographic plates below show photos of acoustic wave propagation obtained in the Royal Institute Lecture Hall (see Ref. [8]). A spherical wave propagates outward from a source. After several reflections (reverberations), the unabsorbed acoustic energy distribution in the room is nearly uniform. Estimate the distance from the source at 0.016 s of the leftmost plate, if the temperature of the lecture hall is 20°C.

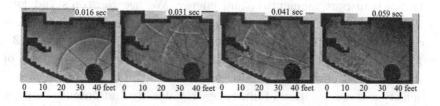

Answer: At 20°C, the speed of sound is 343 m/s. The distance of the wave from the source is $d = (343 \text{ m/s}) \times (0.016 \text{ s}) = 5.5$ m or 18 feet. Note that the shape of the wave spreading outward is that of a spherical shell of ever-increasing surface area.

★ Example 2.24: A speaker connected to a 100-watt amplifier has 1.0% efficiency in converting electric energy to acoustic power. The speaker is located on a platform. If the acoustic power radiating from the speaker is uniform in all directions, what is the intensity of a sound wave at 1.0 m from the speaker? What is the intensity of the sound wave at 2.0 m from the speaker?

Answer: A 100 W amplifier with 1.0% efficiency to sound energy has 1.0 W of acoustic power. The acoustic power spreads out uniformly with directional parameter $Y = 1$, so the intensity at 1 m is $I = \frac{1\,\text{W}}{4\pi(1\,\text{m})^2} = 0.08\frac{\text{W}}{\text{m}^2}$. The intensity at 2 m from the speaker is $I = \frac{1\,\text{W}}{4\pi(2\,\text{m})^2} = 0.02\frac{\text{W}}{\text{m}^2}$, which is $\frac{1}{4}$ of the intensity at 1.0 m. This is the inverse square law.

★ Example 2.25: When the sound wave pressure equals the atmospheric pressure, the sound wave cause "cavitation" of the air, i.e. $\Delta p_{max} = p_{atm}$. What is the intensity of this extreme sound wave?

Answer: $I_{max} = \frac{1}{2\rho v}\Delta p_{max}^2 = \frac{(1.013\times10^5)^2}{2\times400} = 1.3 \times 10^7$ W/m^2. The pressure wave cannot sustain beyond this pressure because there is no air for compression or rarefaction. The ratio to the reference intensity is $I_{max}/I_0 = 1.3 \times 10^{19}$.

VI. Summary

Waves transport momentum and energy. We use Newton's 2nd law, which states that force is the rate of change of momentum (in this case, wave momentum), to derive wave speed in Eqs. (2.2) and (2.3). The wave energy is the sum of the potential and kinetic energies of the motion of participating particles in the medium. The wave energy density is a composition of kinetic energy density and potential energy density in Eq. (2.4). The intensity of a wave is equal to the product of its energy density and its speed in Eq. (2.5). The unit of energy is Joules [J]. The power is the amount of energy per unit time, measured in joules per second or watts [W = J/s]. Therefore, the intensity is the power per unit area, i.e. $I = $ Power/Area, measured in [W/m^2].

A wave is characterized by its frequency f and wavelength λ, and the wave speed is $v = f\lambda$. Waves exhibit phenomena such as reflection, refraction, diffraction, and interference. These phenomena can be explained by Huygens' Principle stating "the wave fronts of a wave become the new sources of new disturbance as the wave travels in media." When an observer and a sound source are in relative motion, the observed frequency differs from the source frequency. This is the Doppler Effect. When a sound source moves faster than the wave speed, the wave fronts pile up and produce a shock wave that can cause sonic boom to observers.

Wave reflection from a fixed-end changes its phase by 180° while wave reflection from a free end does not change phase. Interference between wave and its reflection from the boundary of a physical system can produce resonance, called a standing wave. At certain standing wave frequencies, the wave intensity is highly enhanced, which is the resonance or standing wave. Most physical systems exhibit natural resonance phenomenon. Resonance is the basic design principle of all musical instruments, such as the string, brass and wind instruments. Resonance also plays an important role in human speech and singing. Each resonance is characterized by its resonance frequency f_0, bandwidth Δf, and quality-factor $Q = f_0/\Delta f$.

Sound wave energy density and intensity depend on frequency, density of the medium, wave speed, pressure and displacement amplitudes, shown in Eqs. (2.18) to (2.19). The acoustic impedance is a characteristic property of the transmission medium. Wave reflection occurs when the wave encounter a boundary between two media with different acoustic impedance. The acoustic impedance is also an important design characteristic in musical instruments. The intensity of a sound wave from a point source obeys the inverse square law given by Eq. (2.23), and this is referred to as the "direct sound." The sound energy in a room also needs to include "reverberation," which is the unabsorbed sound energy reflected from walls of the room.

VII.   Homework 02

1. When the incident and reflected waves overlap, _____ may be _____ (leading to a large wave motion) or _____ (leading to a smaller wave motion).

2. _____ is the sensation of high or low of a sound that is mainly determined by the frequency of the sound.

3. Use Eq. (2.7) with $M = 0.02896$ kg/mol and $\gamma = 1.4$ for the air temperature at $T = 273$ K or $0°$C to verify the speed of sound wave in air is 331 m/s. Use $T = 293$ K or $20°$C to verify the speed of sound wave in air is 343 m/s.

4. A violin E-string has a mass per unit length 3.9×10^{-4} kg/m. The tension to produce a transverse wave speed 430 m/s is _____ N.

5. A violin A-string has a mass per unit length 6.4×10^{-4} kg/m. Tension applied to the string is 51 N. The speed of the transverse wave traveling along the string is _____ m/s.

6. The elastic modulus of steel is $E = 16.0 \times 10^{10}$ Pa, and its $\rho = 7860$ kg/m^3. The speed of sound wave in steel is _____ m/s.

7. (choose one) When there is a temperature inversion, the temperature close to the ground is (colder/warmer), and the temperature higher up is (colder/warmer). As a result, the speed of sound high from the ground is (slower/faster) than the speed close to the ground, and the refraction of sound wave bends back towards the ground.

8. Beats result from the addition of two pure tone waves at small difference in frequencies with similar amplitudes. The amplitude of the summed wave varies with the _____ frequency, which is the difference between two frequencies.

9. A pinna has a size of 4.0 cm. Diffraction effects from the pinna are important for frequencies below approximately _____ Hz for sound wave at $20°$C.

10. During a hot thunderstorm, lightning strikes and you hear the thunderclap 3.0 s after the flash. At an air temperature of $86°$F, the

speed of sound is _____ m/s and the distance from you to the lightning bolt is _____ km. Does your number agree with the saying of "3-sec-km" or "5-sec-mile" rule of thumb?

11. The minimum ultrasonic sound frequency, at diffraction limit, needed to resolve features on a fetus of size less than 0.5 mm is _____ MHz (Assuming the speed of sound in human tissue is 1500 m/s. The intensity in this procedure is generally less than 1 W/cm^2 to avoid tissue heating and cavitation effects in the object under examination.)

12. Two waves of the same frequency have pressure amplitudes 2.0 N/m^2 and 4.0 N/m^2. In traversing the same medium, they undergo interference. If the interference is constructive, then the resulting pressure amplitude is _____ N/m^2. If the interference is destructive, then the resulting amplitude is _____ N/m^2.

13. Two identical loudspeakers emitting the same pure tones ($f = 343$ Hz) are facing each other at a distance of 10.0 m. If the speed of sound is 343 m/s, the wavelength $\lambda =$ _____ m. (a) An observer positioned at the middle between the two speakers hears _____ interference. (b) If the observer moves 0.25 m towards one of the speakers, the sound waves from two speakers at the observer's new position will be _____ interference.

14. You are tuning a piano using middle A tuning fork with a frequency of 440 Hz. You hit the middle A key of the piano and hear beats at a frequency of 2 Hz. That piano key has a frequency of either $f_1 =$ _____ Hz or $f_2 =$ _____ Hz (where $f_1 < f_2$).

15. Two pure tones of frequency f_1 and f_2 are played simultaneously. The superposition (sum) of the two tones results in a new tone of frequency of $f = 504$ Hz with an amplitude that varies with a

frequency of 4 Hz. The two frequencies are $f_1 = $ _____ Hz
and $f_2 = $ _____ Hz (where $f_1 < f_2$).

16. The _____ effect is the apparent shift in frequency
    observed when the source and the observer are in relative motion.
    When they move towards each other, the frequency appears
    _____; when they move away from each other, the fre-
    quency appears _____.

17. The average and the rms value of the following 10 numbers (3.5,
    -3.5, 2.2, -2.2, 2.5, -2.5, 1.2, -1.2, 0.6, and -0.6) are respectively
    _____ and _____ (see Appendix B).

18. When a sound source, such as a supersonic plane, moves faster
    than the speed of sound, the wave fronts pile up and produce a
    cone-shaped _____ _____ (two words) that creates a
    _____ _____ as it passes a stationary observer.

19. The frequency of an ambulance siren at rest is 1000 Hz and the
    speed of sound in air is 770 mi/hr. You are standing at the
    side of the road as the ambulance is moving with a speed of
    45 mi/hr towards you and then passes you. The frequency of
    the sound you heard as the ambulance is moving towards you is
    _____ Hz. The frequency of the sound you hear as the
    ambulance is moving away from you is _____ Hz.

20. A _____ wave is the stationary wavelike pattern
    that results from the interference of two or more traveling
    waves. Standing waves have regions of minimum amplitude
    called _____ and regions of maximum amplitude called
    _____.

21. A violin G-string has a mass per unit length 2.6×10^{-3} kg/m. The
    tension on the violin G-string is 44 N. The speed of wave on the
    string is _____ m/s.

22. A string fixed at both ends has a length $L = 75$ cm. The wavelengths of the fundamentals and of the first two harmonics are; $\lambda_1 = $ _____ m, $\lambda_2 = $ _____ m, $\lambda_3 = $ _____ m.

23. A string fixed at both ends has a length $L = 75$ cm. If the velocity of transverse wave is 125 m/s, the fundamental frequency is $f_1 = $ _____ Hz.

24. A nylon guitar string, 65 cm long, has a mass of 5.8×10^{-4} kg and a tension of 51 N. The speed of the wave on the string is _____ m/s.

25. A nylon guitar string, 65 cm long, has a mass of 5.8×10^{-4} kg and a tension of 51 N. The wavelengths of the 3 lowest resonance modes are _____ & _____ & _____ cm.

26. A nylon guitar string, 65 cm long, has a mass of 5.8×10^{-4} kg and a tension of 51 N. The frequencies of the 3 lowest resonance modes are _____ & _____ & _____ Hz.

27. A certain string is formed to have a frequency difference between the 3rd and 5th harmonic of 250 Hz. The frequency of the fundamental mode $f_1 = $ _____ Hz.

28. The simplest standing wave, where a string vibrates in one loop at its lowest frequency, is the _____ . The other standing waves are _____ .

29. A steel violin string vibrates at a fundamental frequency of 440 Hz. The length of the string is 0.33 m and the mass of the string of this section is 0.36×10^{-3} kg. The wavelength of the fundamental mode on the string is _____ m. The speed of the transverse wave on the string is _____ m/s. The tension is _____ N.

30. Sound waves move through air at 20°C with a speed of 343 m/s. Model the Ear canal as a closed tube at 1 inch, the fundamental resonance frequency is _____ Hz.

31. The trombone behaves like a pipe that is open at both ends. The lowest note on a 10-ft-long trombone is _____ Hz at 20°C.

32. The fundamental of a closed organ pipe is 261.63 Hz (middle C) at 20°C. The length of the pipe is _____ m.

33. The second harmonic of an open organ pipe is 261.63 Hz at 20°C. Its length is _____ m.

34. A 16-ft open organ pipe (both ends open) is operating in air at 20°C. The frequencies of the first three modes are _____ Hz, _____ Hz and _____ Hz.

35. A 16-ft open organ pipe is operating in air at 20°C. If the air is replaced by helium at 20°C with $v = 1005$ m/s. The new fundamental frequency is _____ Hz.

36. A 16-ft closed organ pipe (one end open and other end closed) is operating in air at 20°C. The frequencies of the first three modes are _____ Hz, _____ Hz and _____ Hz.

37. A 16-ft closed organ pipe (one end open and other end closed) is operating in air at 20°C. If the air is replaced by helium at 20°C with $v = 1005$ m/s. The new fundamental frequency is _____ Hz.

38. Two identical organ pipes. One filled with regular air at 20°C has a frequency of 500 Hz. If the other pipe is filled with Helium gas at 20°C, its frequency is _____ Hz. The speed of the sound wave in He gas at 20°C is 1005 m/s.

39. Two identical organ pipes, both at 20°C, have frequencies of 500 Hz. If the temperature of one pipe is raised to 33°C, its new frequency is _____ Hz.

40. The length of a closed organ pipe is operating in air at 20°C to produce a fundamental frequency of 20 Hz is _____ m.

41. We model the outer ear canal as a cylindrical pipe 2.6 cm long that is closed at one end by the eardrum. The resonant (fundamental) frequency for this closed pipe is = _____ Hz.

42. We model the outer ear canal as a cylindrical pipe 2.6 cm long that is closed at one end by the eardrum. At the eardrum end, is it a pressure wave antinode or nodal point?

43. The vibration spectrum gives the frequency and amplitude of each mode of vibration excited. The spectrum of a plucked guitar string is shown below:

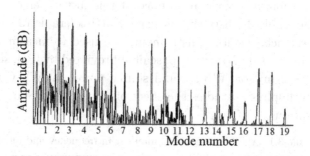

The lowest frequency is the _____ and all higher frequencies are _____ .

44. Find the resonance frequency, bandwidth, and the Q-factor of the resonance curve shown in Fig. 2.13.

45. At the typical conversation between people, the rms sound pressure is $\Delta p_{rms} = 0.020$ N/m^2. Assuming that the sound wave is 1000 Hz, what is the amplitude of the oscillation of the air molecules?

46. The sound intensity at 6 m from the speaker in a rock concert is 1.0 W/m^2. (a) What is the sound intensity at 3 m from the speaker? (b) What is the power of the speaker? (Assuming uniform spreading of the sphere with directivity $Y = 1$.)

Chapter 3

Sound Levels, Loudness and Psychoacoustics

Three subjective attributes of our perception of sound are loudness, pitch, and timbre. The corresponding physical parameters are the pressure amplitude, frequency, and sound spectrum and pulse envelope. Pulse duration also plays an important role in the perceived loudness and pitch. Table 3.1 lists the sensitivity and tolerable range of human ears. In science, scientific instruments are used to measure the sound pressure in units of $[N/m^2]$ or [pascal] abbreviated by [Pa], sound intensity in $[watt/m^2]$ or $[W/m^2]$, and source power in [W]. On the other hand, psychophysics studies the effect of sound pressure on our subjective sensations.

Table 3.1: Audible range of human ear in frequency and intensity.

Frequency	Intensity	Intensity level	Pressure
20–20,000 (Hz)	10^{-12}–10 (W/m^2)	0–130 (dB)	2×10^{-5}–60 (N/m^2)

To accommodate the incredible sensitivity range of 12 orders of magnitude in intensity shown in Table 3.1, physicists and engineers use the decibel unit or [dB]. The decibel scale is the sound *"level."* Since our ears' perception of loudness level depends also on frequency, we define the psychoacoustic equal loudness "level" as the [phon], which depends on both the sound level and frequency. Finally, what is the meaning of "twice as loud"? Psychoacoustic surveys lead to the unit of "loudness" in [sone]. Topics in this chapter include (1) the human hearing system, including outer/middle/inner ear and the central nervous system (CNS); (2) psychoacoustic laws; (3) sound intensity level, sound pressure level in decibels, and loudness level; (4) loudness of pure

72

and complex tones; (5) critical bands, loudness of combined sounds and multiple sources, and musical dynamics and loudness; (6) masking; and (7) effects of binaural hearing.

I. Human Hearing System

We conveniently divide our ear into three sections: outer ear, middle ear, and inner ear, schematically shown in Fig. 3.1. The outer ear is composed of the pinna and ear canal. The length of our ear canal is about 2.5 cm, which enhances standing wave resonances at frequencies 3.5 kHz and 10.5 kHz with a gain of about 5 dB (see Example 2.16).

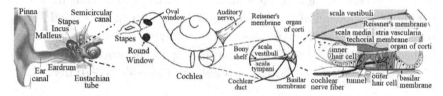

Figure 3.1: Anatomy of the human ear, divided into the outer, middle and inner ears. The inner ear drawing shows the cochlea and its cross-section, revealing the basilar and Reissner's membranes separating the canals into chambers. The expanded view shows the cochlear duct, the organ of Corti, and the inner and outer hair cells. Sound waves induce vibrations in the oval canal, setting up waves in the cochlea, producing nerve signals transmitted to the brain.

The middle ear is composed of eardrum and the ossicles (malleus or hammer, incus or anvil, and stapes or stirrup), 6 ligaments (3 on the malleus, 2 on the incus and 1 on the stapes) and 2 muscles (tensor tympanic and tensor stapedius). The ossicles in the middle ear couple the sound vibration from the air to fluid vibration in the cochlea of the inner ear. The ossicles act as a lever with a force advantage of about 1.3–1.5. The ratio of the areas between the tympanic membrane (area $\approx$ 60–70 mm^2) and the oval window (area $\approx 3 \sim 4$ mm^2) gives another factor of $15 \sim 20$ amplification on sound pressure. The combined amplification of acoustic pressure is about 17–25, i.e. a gain of about 25 dB. The muscles in the middle ear provide some protection of the inner ear through acoustic reflex. The Eustachian tube links the middle ear to the nasopharynx, which consists of the upper throat and the back of the nasal cavity. When we swallow, the Eustachian tube opens for pressure balance and fluid drainage out of the middle ear.

The inner ear consists of semicircular canals and the cochlea. The semi-circular canals have three orientations that provide equilibrium and balance in three-dimensional space. The cochlea is a snail-shaped canal encased in the temporal bone of the skull. In humans, the snail shell completes about $2\frac{1}{2}$ to $2\frac{3}{4}$ turns, and its uncoiled length is about 35 mm. The cross-section of the cochlea has 3 chambers along most of its length: the scala vestibule, the scala tympani, and the scala media. The scala vestibule connects to the scala tympani at the helicotrema, schematically shown in Fig. 3.2(a). They are filled with perilymph, which is a liquid similar in composition to the fluid surrounding body cells. The cochlear membrane is an elastic partition that runs from the beginning to the end of the cochlea, splitting it into an upper and lower part. The cochlear duct is filled with endolymph, a liquid having a higher concentration of K^+ ion, maintaining an endocochlear potential of +80 mV, while a cell body has a potential of -55 to -70 mV.

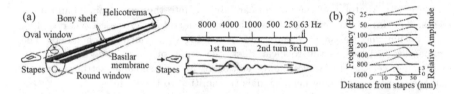

Figure 3.2: (a) The anatomy of the cochlea at the basilar membrane. Vibration of the stapes sets up waves on the membrane. The position of maximum vibration amplitude is a function of frequency. High frequencies set up vibrations close to the oval window, while low frequencies set up oscillations at the helicotrema; (b) Relative amplitude of a traveling wave oscillation on the basilar membrane as excited by sound waves at various frequencies (historical data by Békésy).

The vibration of the stapes transmits sound pressure waves through the oval window into the scala vestibuli canals and sets up a traveling wave vibration on the cochlear membrane. As the hair cells in the cochlear duct move up and down, microscopic hair-like projections (known as stereocilia) perched on top of the hair cells bump against an overlying structure and bend. The bending causes pore-like channels at the tips of the stereocilia to open up; and K^+ ions rush into the cell, creating an electrical signal transmitted to our central nerve system (CNS). This creates an electric neuron pulse signal. In 1939, Hallowell Davis and Robert Galambos made the first recordings of the electrical activity of individual neurons.[1] The auditory nerve (AN) or cochlear nerve transmits electrical pulses that provide temporal information to our auditory brainstem and auditory cortex.

The basilar membrane in the cochlea is a frequency analyzer, as schematically shown in Fig. 3.2(a) with a wave traveling along the basilar membrane. There are about 12,000 outer hair cells in 3 rolls and 3,500 inner hair cells in 1 roll along the length of the cochlea. The Basilar membrane is narrow (0.08–0.16 mm) and stiff at the base, and wider (0.42–0.65 mm) and less stiff at the apex of the cochlea, known as the *helicotrema*. The hair cells, particularly the inner hair cells, convert mechanical vibrations of the basilar membrane into electrical signals transmitted through neurons to the CNS. In 1924–1946, G.v. Békésy carried out systematic experiments and observed traveling wave on the basilar membrane.[2] Figure 3.2(b), observed by Békésy, shows the frequency response of the cochlea of a dead animal to a pure sinusoidal wave. The response curve (or the tuning curve) is much sharper in live animals, and the response curve will broaden in a few hours after death. High-frequency disturbances excite the basilar membrane close to the oval window. In 2006, Manoussaki, Dimitriadis, and Chadwick showed that the curved geometry of the cochlea enhances low frequency standing wave amplitude by a factor of about 10 on the basilar membrane.[3]

Physiological research discovered the existence of *characteristic frequency* (CF) for each hearing hair cell on the cochlear membrane and their associated auditory neurons, where the CF is the frequency to which the cell has a maximum response.[4] The Greenwood CF formula, which fits human cochlea, is

$$CF = A(2^{\alpha x} - k), \qquad (3.1)$$

---

[1] H. Davis and R. Galambos, *J. Neurophysiol.*, **6**, 39–58 (1943).

[2] See http://www.annualreviews.org/doi/pdf/10.1146/annurev.ph.36.030174. 000245; In 1961, G.v. Békésy received Nobel Prize in Physiology and Medicine.

[3] Manoussaki, Dimitriadis, and Chadwick, *Phys. Rev. Lett.*, **96**, 088701 (2006).

[4] D.D. Greenwood, *JASA*, **87**, 2592 (1990).

where $\alpha \approx 7.0$, the parameter $x \in [0, 1]$ is the normalized distance from the apex of the cochlear membrane, $k \approx 0.88$, and $A \approx 165.4$ Hz. Figure 3.3(a) shows the Greenwood CF curve vs the distance from the apex for a cochlear length of 35 mm. The CF at the mid-point of the cochlear membrane is 2 kHz. For $x > 0.3$, i.e. 30% of the cochlear membrane from the apex, the constant k is small compared to the exponential term in CF formula. Each octave will occupy equal length along the cochlear membrane, with $\Delta x \approx 1/\alpha \approx 14\%$. The 70% of the cochlear length on the "base side" of cochlea covers about 5 octaves, i.e. $0.7/0.14 = 5$. Since human hearing is about 10 octaves, the remaining 30% of the cochlear membrane at the apex end covers 5 octaves.

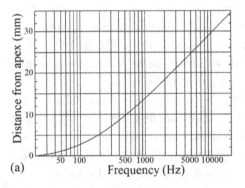 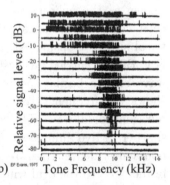

Figure 3.3: (a) Greenwood CF formula of the human cochlea vs the distance $d = x \times (35 \text{ mm})$ from the apex, for a person with a cochlear membrane 35 mm in length. (b) The neuro-tuning curve for a single cochlear AN of a guinea pig vs frequency at relative sound level (E.F. Evans).

Figure 3.3(b) shows the tuning curve of a single auditory neuron fiber with CF at 10 kHz of guinea pig.[5] A neuron fiber tuned to a certain characteristic frequency can respond to a louder sound at lower frequencies, but not to sound at higher frequencies. In other words, higher-frequency sound cannot reach far enough along the cochlear membrane to excite the neuron fiber that has a lower characteristic frequency. This property of the tuning curve can explain sound masking. The fact that "our frequency analysis is associated with location along the cochlear

---

[5]E.F. Evans, *J. Physiology*, **226**, 263 (1972); DOI: 10.1113/jphysiol.1972.sp009984

membrane and each neuron fiber has its own CF" supports the *place theory of hearing*.[6]

Experimental measurements seem to indicate that the response curve of each individual neuron fiber resembles the *gamma-tone filter function*:[7]

$$g(t) = at^{n-1}e^{-2\pi bt}\cos\left(2\pi ft + \varphi\right), \tag{3.2}$$

where f is the characteristic frequency, φ is the phase of the carrier, a is the amplitude, n is the order of the filter function, and b is the filter bandwidth. Scientists use the Gamma-tone-filter-function to model the impulse response of auditory filters for auditory system.

In addition to the sharp frequency tuning in cochlear vibrations and in auditory nerve fibers, the resulting frequency analysis is transmitted by the auditory nerve by a series of pulses to the brain. The auditory nerve pulse retains the time structure of the sound sources. For the signal of a nerve fiber for a single tone, "the nerve fiber may not fire at the peak of every period, but it rarely fires at any other time."[8] This fact supports the *temporal theory of hearing*.

There are about 30,000 auditory nerve fibers in each inner ear that transmit the auditory nerve pulses through the chain of structures including the cochlear nucleus, superior olive, lateral lemniscus, inferior colliculus, and medial geniculate body to the auditory cortex on both sides of auditory neural network. Combination and comparison of signals from both ears occur in some structures in the neural transmission network. These neural signals retain "temporal structure" of the sound source and at the same time are "tonotopically organized in regions" of the structure in the transmission chain. Our hearing system employs both temporal and spatial information to attain relevant perceptions of the environment, such as sound source localization, pitch recognition, etc.

---

[6]A comprehensive review of cochlear membrane mechanism is reported by Luis Robles and Mario A. Ruggero, Mechanics of the mammalian cochlea, *Physiological Reviews*, Vol. 81 no. 3, 1305–1352 (1 July 2001) (see Fig. 18 in http://physrev. physiology.org/content/81/3/1305).

[7]T.J. Goblick, Jr. and R.P. Pfeiffer, *JASA*, **46**, 924 (1969).

[8]See e.g. I. Tasaki, *J. Neurophysiol.*, **17**, 97 (1954).

The mechanical vibration of the cochlear membrane in the inner ear can also cause eardrum to vibrate and generate low-level sound wave, measurable at the ear canal. This low-level eardrum vibration is called an otoacoustic emission (OAE). In 1978, David Kemp demonstrated the phenomenon experimentally based on a prediction by Thomas Gold in 1948.[9] The otoacoustic emissions may arise through a number of different cellular and mechanical mechanisms within the inner ear. Studies have shown that OAEs disappear in damaged inner ears; thus, OAEs are often used in labs and clinics to gauge the health of the inner ear. The applications of OAE are still an important research topic in medical communities.

II.   Psychoacoustics

Psychophysics is the study of the relationship between stimuli and the corresponding subjective sensations. G.T. Fechner (1801–1887), inspired by Ernst Weber (1795–1878), pioneered in trying to determine the relationship between the stimuli and the perceived sensation in human subjects.

A.   Weber–Fechner Law and Logarithms in Sound and Music

According to the Greenwood formula in Eq. (3.1), the frequency response of our ear is logarithmic. The music scale is logarithmic. The intensity of a sensation and the loudness of the sound intensity level increase logarithmically. Our psychoacoustic sensation on hearing appears to obey the Weber–Fechner law: "our sensation is proportional to the logarithm of stimulus."

B.   Basic Mathematics of Logarithm Function (base-10)

The logarithm (base-10) of a number is the power to which 10 (the base) must be raised to return the number, i.e. $10^{\text{Log}(x)} = x$ for

---

[9]D.T. Kemp, *JASA*, **64**, 1386 (1978); T. Gold, *Proc. Roy. Soc. B*, **135**, 492 (1948).

any $x > 0$. Appendix C provides some basic properties of logarithmic (log) functions. We list some of these properties below.

Log (A × B) = Log A + Log B; Log (A/B) = Log A − Log B; Log A^n = nLog A

The values of the logarithm of some numbers are tabulated below:

x	Log x	x	Log x	x	Log x	x	Log x	x	Log x
1	0	5	0.699	9	0.954	0.001	−3	10	1
2	0.301	6	0.778	10	1	0.01	−2	100	2
3	0.477	7	0.845	11	1.041	0.1	−1	1000	3
4	0.602	8	0.903	12	1.079	1	0	10000	4

III. Sound Levels

Since the dynamic range of our hearing in terms of sound intensity is huge, from 10^{-12} to 1 W/m^2, it is easier to express sound levels in logarithmic scale. Electronic engineers also use the logarithmic function to express power gain, attenuation and amplification. This section defines the sound power level, sound intensity level, and sound pressure level in decibels or [dB]. A "decibel" is $\frac{1}{10}$ of a "bel," which is a unit first introduced by Alexander Graham Bell.

A. Power "Level"

The definition of power "level" for any power source is

$$L_W \equiv 10 \log \frac{W}{W_0} \text{ with reference power } W_0 = 10^{-12} \text{ W}. \qquad (3.3)$$

The "decibel (level) gain/reduction" of two power levels is $\Delta L = L_2 - L_1 = 10 \log \frac{W_2}{W_1}$, where $\frac{W_2}{W_1}$ is the power ratio. Note that $\Delta L > 0$ corresponds to $\frac{W_2}{W_1} > 1$; and $\Delta L < 0$ for $\frac{W_2}{W_1} < 1$.

★ Example 3.1: What is the Sound Power Level of a speaker that radiates 0.1 W?

Answer: The power level of the speaker is $L_W = 10 \log \frac{0.1}{10^{-12}} =$ 110 dB.

★ Example 3.2: Find the decibel gain of an amplifier that produces an output of 10 W at 0.1 W input.

Answer: The decibel gain of the amplifier is $\Delta L_w = 10 \log \frac{W_2}{W_1} = 10 \log \frac{10}{0.1} = 10 \log 100 = 20$ dB.

★ Example 3.3: What is the decibel gain (or power level gain) when the power gain is 400?

Answer: The power gain is $\frac{W_2}{W_1} = 400$, and the decibel gain is $\Delta L_w = 10 \log 400 = 26$ dB.

★ Example 3.4: An amplifier has a decibel gain of 30 dB, what is the power gain?

Answer: The decibel gain is $\Delta L_w = 10 \log \frac{W_2}{W_1} = 30$ dB, we find $\frac{W_2}{W_1} = 10^3$.

B. Sound Intensity Level (SIL) or "Sound Level" L_I

Intensity is power per unit area. The sound waves that reach our ear canals are relevant to our hearing. The sound intensity level that characterizes the sound intensity in acoustics is

$$L_I \equiv 10 \log \frac{I}{I_0}, \tag{3.4}$$

where the "threshold intensity" of hearing $I_0 = 10^{-12}$ W/m^2 is chosen for the reference intensity. Table 3.2 lists the typical sound intensity and sound level of some events. Portable music players' sound intensity level can be as high as 120 dB. The sound level at the front row of a rock concert can cause damage to our ears. In 2006, the iPod maker Apple introduced software for users to set loudness limits after a lawsuit was filed against the company for contributing to hearing loss. In 2009, the European Union introduced a regulation requiring that all MP3 players' default maximum volume setting is 85 dB.

★ Example 3.5: What is the sound intensity level at a location where $I = 10^{-4}$ W/m^2?

Answer: The SIL is $L_I = 10 \log \frac{I}{I_0} = 10 \log \frac{10^{-4}}{10^{-12}} = 80$ dB.

Table 3.2: Some typical sound level of some commonly occurring events.

Sources	Intensity	Sound Level
Threshold of Hearing (TOH)	1×10^{-12} W/m^2	0 dB
Whisper at 1 m	1×10^{-10} W/m^2	20 dB
Normal Conversation at 1 m	1×10^{-6} W/m^2	60 dB
Busy Street Traffic	1×10^{-5} W/m^2	70 dB
Vacuum Cleaner, heavy traffic	1×10^{-4} W/m^2	80 dB
Large Orchestra	6.3×10^{-3} W/m^2	98 dB
Front Rows of Rock Concert	1×10^{-1} W/m^2	110 dB
Threshold of Pain	1×10^{1} W/m^2	130 dB
Military Jet Takeoff	1×10^{2} W/m^2	140 dB
Instant Perforation of Eardrum	1×10^{4} W/m^2	160 dB

★ Example 3.6: If a trombone bell has an area 0.1 m^2 and the sound power radiated from the bell is 1.5 W, what is the sound intensity level at the bell?

Answer: The intensity of the trombone at the bell is $I = \frac{\text{Power}}{\text{Area}} = \frac{1.5}{0.1} = 15 \frac{\text{W}}{\text{m}^2}$. The sound intensity level is $L_I = 10 \log \frac{I}{I_0} = 10 \log \frac{15}{10^{-12}} = 132$ dB.

★ Example 3.7: The sound power level of a loudspeaker is 140 dB. Find the distance from the speaker at which the sound intensity level is 90 dB (assuming uniform spread and no absorption).

Answer: First, 140 dB = 10 Log (W/W_0), whence the sound power of the speaker is $W = 100$ W. At a location where SIL = 90 dB = 10 Log (I/I_0), the sound intensity is $I = 10^{-3}$ W/m^2. Using the inverse square law: 10^{-3} W/m$^2 = \frac{100\,\text{W}}{4\pi r^2}$, we find $r = 89$ m.

★ Example 3.8: The SIL of a running motor on a concrete floor is 95 dB at a distance of 1 m. Find the sound power of the motor, and the Sound Power Level of the motor.

Answer: For SIL 95 dB, we use 95 dB = 10 Log (I/I_0) to obtain $I = 10^{9.5} I_0 = 0.00316$ W/m^2. On a concrete floor, the sound intensity obeys the inverse square law with directivity $Y = 2$ i.e. $I = 2\frac{W}{4\pi r^2}$. The acoustic power of the motor $W = 2\pi r^2 I = 2\pi (1\,\text{m})^2 \times 0.00316\frac{W}{m^2} = 0.020$ [W].

C.   Sound Pressure Level (SPL) L_p

The average sound intensity is $I = p^2/\rho v$, where p is the rms sound pressure, $\rho = 1.293$ kg/m^3 is the density of the air and $v = 331$ m/s is the speed of sound in air. The acoustic impedance of the air ρv is **set** at $\rho v = 400$ kg/(m^{2}s). The reference sound intensity is $I_0 = p_0^2/400 = 10^{-12}$ W/m^2. We find the reference rms pressure $p_0 = 2 \times 10^{-5}$ N/m^2. The sound pressure level (SPL), defined as

$$L_p = 10\log\left(\frac{p}{p_0}\right)^2 = 20\log\frac{p}{p_0}, \quad \text{with} \quad p_0 = 2.0 \times 10^{-5}\,\text{N/m}^2, \quad (3.5)$$

is identical to the SIL of Eq. (3.4). Thus, SIL and SPL are identical. They are simply *"sound level."*

★ Example 3.9: What is the SPL for a sound rms pressure of $p = 0.001$ N/m^2 and 0.002 N/m^2?

Answer: Using Eq. (3.5) with $p = 0.001$ Pa, we find $L_p = 20\log\frac{0.001}{2\times10^{-5}} = 34$ dB; similarly, using $p = 0.002$ Pa, we find $L_p = 20\log\frac{0.002}{2\times10^{-5}} = 40$ dB. When pressure increases by a factor of 2, the sound level increases by 6 dB.

★ Example 3.10: How much force does a sound wave at the $L_p = 120$ dB exert on an eardrum having diameter of 7 mm?

Answer: 120 dB $= 20\log(p/p_0)$; so $p = 10^{120/20} \times p_0 = 20$ N/m^2. The magnitude of the rms force on the eardrum is $F = p \times$ Area $= 20\frac{N}{m^2} \times \pi\left(\frac{0.007}{2}\,\text{m}\right)^2 = 0.00077$ N. This force seems small, but a sound at the loudness level can damage our ear.

D.   Sound Intensity Level of Multiple Sources

For independent and uncorrelated sound sources, we add sound intensities to obtain the combined excitation of the medium. The combined intensity of several (incoherent) sources is the sum of individual intensities, i.e. $I = I_1 + I_2 + \cdots$ [see Eq. (2.13)].

★ Example 3.11: The SPL of one sound source is 60 dB. Find the SPL of the combined 2 identical incoherent sound sources at 60 dB each?

Answer: Typical sound sources are incoherent. The combined intensities add. Thus the intensity of 2 sources with the same SPL is $I_2 = 2I_1$. The SPL of 2 sources is

$$L_2 = 10\log\frac{I_2}{I_0} = 10\log\frac{2I_1}{I_0} = 10\left(\log 2 + \log\frac{I_1}{I_0}\right) = 3+60 = 63\text{ dB}.$$

The sound level increases by 3 dB when doubling the sound intensity. The combined sound level of two incoherent (uncorrelated) sound sources at 40 dB each is 43 dB. Removing one sound source from two identical incoherent (uncorrelated) sound sources at a combined sound level of 60 dB is 57 dB.

★ Example 3.12: A single violin plays at an SPL of 50 dB. What is the SPL of 3 violins playing at the same SPL?

Answer: Let the intensity of one violin be I_1, where $10\log(I_1/I_0) = 50$ dB. When the three violins play at equal sound level, the combined sound intensity is $I_3 = 3I_1$. The SPL of 3 violins is

$$10\log(3I_1/I_0) = 10\log(3) + 10\log(I_1/I_0) = 4.8 + 50 = 54.8\text{ dB}.$$

★ Example 3.13: Three are three sound sources at SPL of 50, 60 and 70 dB. What is the SPL of the combined sound?

Answer: The intensities for 50, 60, and 70 dB sound sources are 10^{-7}, 10^{-6}, and 10^{-5} W/m^2, respectively. The combined sound intensity is $I_1 + I_2 + I_3 = 10^{-7} + 10^{-6} + 10^{-5} = 1.11 \times 10^{-5}$ W/m^2. Therefore, the SPL of the combined sound is

$$L_p = 10\log\frac{I_1 + I_2 + I_3}{I_0} = 10\log\frac{1.11 \times 10^{-5}}{10^{-12}} = 70.5\text{ dB}.$$

★ Example 3.14: What is the combined SPL of two sound sources with SPL at 50 dB and 53 dB?

Answer: The intensities of SPL 50 and 53 dB are respectively $I_1 = 10^{-7}$ W/m^2, $I_2 = 2 \times 10^{-7}$ W/m^2. The combined sound intensity is $I_1 + I_2 = 3 \times 10^{-7}$ W/m^2. The SPL of the combined sound is

$$L_p = 10\log\frac{3 \times 10^{-7}}{10^{-12}} = 54.8\text{ dB}.$$

IV.   Loudness Level L_L, Equal Loudness Level and Loudness

Sound level is a property of sound source intensity. The loudness level is the psychoacoustic effect of the sound level on a listener. The sensitivity of the human ear depends on the sound frequency. In 1933, H. Fletcher and W.A. Munson carried out psychoacoustic experiments to set an "equal loudness level (L_L)" as a function of frequency. Figure 3.4 shows the equal loudness level contour L_L vs frequency, revised by the International Standard Organization (ISO 226:2003).

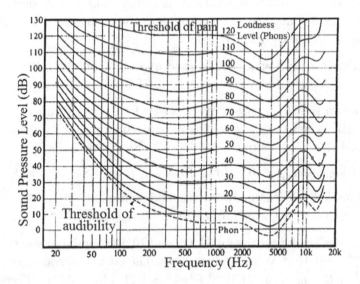

Figure 3.4: ISO equal loudness contour or the revised Fletcher–Munson curve, based on a review of population survey made in many countries. The equal loudness level curve shows the maximum sensitivity of human ear peak at 3,500–4,000 Hz, which is the first resonance frequency of ear canal, and again at about 13 kHz, the second resonance.

The unit of the "loudness level" is the "phon," defined as the dB sound level at 1,000 Hz. For example, 40 dB SPL at 1,000 Hz is 40 phon, etc. The 40 dB sound level at 1,000 Hz has equal perceived loudness level as a 52 dB sound at 100 Hz, and the entire equal loudness level curve is 40 phon. The "phon," NOT widely used, is defined in the equal loudness level contour by the ISO 226:2003, shown in Fig. 3.4.

The sound intensity or pressure "level" in dB and the loudness "level" in phon are not psychoacoustic "loudness." What is the concept of "twice as loud"? How do we perceive a sound that is twice as loud as another is?

Some psychoacoustic experiments find that a "10-phon" (or 10-dB) increment increases the "loudness" by a factor of 2. The International Standard Organization (ISO) recommends that a sound is "twice as loud" when the sound "level" increases by 10 phon. The unit for this loudness is the "sone," defined as the loudness of a 1,000 Hz pure tone at a "sound level" of 40 phon. Table 3.3 lists "loudness levels" and their associated "loudness" for a pure tone.

Table 3.3: The loudness level (phon) vs the loudness (sone) for a pure tone.

Loudness level (phon)	40	50	60	70	80	90	100	110	120
Loudness (sone)	1	2	4	8	16	32	64	128	256

Using the symbol S for the loudness in sone and L_L for the loudness level in phon, we find

$$S = 2^{(L_L-40)/10} \text{ [sone], or } L_L = 40 + 10 \times \text{Log}_2 S \text{ [phon]}. \qquad (3.6)$$

★ Example 3.15: (Reading of the equal loudness level curve) Find the loudness level of a pure tone at 100 Hz and SPL 30 dB.

Answer: The ISO 226:2003 curve passing through the point 100 Hz and SPL 30 dB is the 10 phon line. Thus, a 100 Hz pure tone at SPL 30 dB gives 10 phon.

★ Example 3.16: What is the SPL for a pure tone at 3,000 Hz that gives a loudness level of 70 phon?

Answer: The 70-phon curve of ISO 226:2003 passes through 3,000 Hz at SPL 62 dB. Thus, 62 dB sound level at 3,000 Hz gives 70 phon.

★ Example 3.17: What is the loudness level and loudness of a 3,000 Hz pure tone at $L_p = 70$ dB?

Answer: The equal loudness curve passing through 3,000 Hz and 70 dB location is 80 phon. The loudness is $2^{(80-40)/10} = 2^4 = 16$ sone.

★ Example 3.18: What is the SPL to produce loudness 2 sone for a
pure tone 100 Hz?

Answer: For a loudness 2 sone, the loudness level is 50 phon. The
50 phon line of equal loudness (ISO 226-2003 curve) passes through
100 Hz at SPL 60 dB.

For a pure tone, the loudness "**sone**" increases by a factor of 2
when the loudness level increases by 10 phon or 10 dB. Using $L_L = 20 \log (p/p_0)$, we find $S = Cp^{0.6} \sim I^{0.3}$, where C depends on frequency, p is the rms sound pressure, and p_0 is the reference pressure.
At 1 kHz, one finds $C = 42$ sone for the rms pressure p in the unit
of [Pa]. Some psychoacoustic experiments found $S \sim p^{0.5} \sim I^{0.25}$ for a
wideband signal. Similarly, some researchers found the doubling of loudness with 6-dB increases in loudness level, i.e. $S = 2^{(L_L-40)/6}$. In this
case, one finds $S = Kp$, where K also depends on frequency. Although
the idea of a power law had been suggested by researchers in the 19th
century, Stanley Smith Stevens (1906–1973) is credited with reviving
the law in 1957. Instead of the Weber–Fechner's psychoacoustics law
of *Sensation* $\sim \log(I)$, Steven's power law is *Sensation* $= kI^a$, where I
is the magnitude of the physical stimulus, *Sensation* is the subjective
magnitude of the sensation evoked by the stimulus, a is an exponent
that depends on the type of stimulation, and proportional constant k
is a complicated function of frequency. The "loudness" (in sone) seems
to obey the power law.

A.   Sound Level Meters

Our perception of loudness depends sensitively on frequency. There are
three common weighting filter networks used in the measuring system,
designated as A, B and C filters, to reflect the sensitivity of human
ears. The resulting sound levels, designated as dBA, dBB, and dBC, are
equivalent to 40 phon, 70 phon and 100 phon of the ISO 226:2003 equal
loudness level curve. Figure 3.5 shows the frequency response of these
filters. The frequency response of the C-weighting network is nearly
flat. The A-weighting network introduces low-frequency gain roll-off to
reflect the frequency response of the human ear at low sound pressure

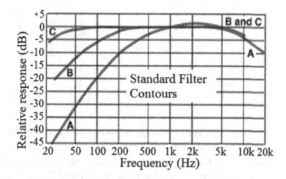

Figure 3.5: Filter function of ABC-weighted sound level meters. The weighting function of dB(A/B/C) resembles the 40/70/100 phon equal loudness level in ISO 226:2003.

levels. The measurement of sound levels using the A-weighting factor has the unit of dBA or dB(A).

B. Musical Dynamics and Loudness

Music dynamics is the variation in loudness between notes and phrases. The symbols are fortissimo (ff); forte (f); mezzo forte (mf); mezzo piano (mp); piano (p); and pianissimo (pp). Table 3.4 lists the corresponding A-weighted sound level. We note that the typical sound pressure level in speech is about 30 to 80 dBA.

Table 3.4: Estimation of the loudness (in sone) for the A-weighted sound level.

Dynamic level	ppp	pp	p	mp	mf	f	ff	fff
dB (A-weighted)	30	40	50	60	70	80	90	100
Sone	$\frac{1}{2}$	1	2	4	8	16	32	64

V. Critical Band, Loudness of Complex Tones and Combined Sounds

One defines the "spectral" intensity level (L_{spectral}) as the sound intensity level at 1 Hz bandwidth [see the top plot Fig. 3.6(a)]. A broadband source is a sound source characterized by a center frequency f_c and a

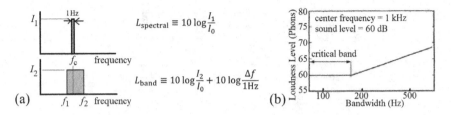

Figure 3.6: (a) Schematic drawing of a narrow-band and broadband sound source centered at the center frequency f_c. (b) Schematic drawing of the definition of Fletcher's psychoacoustic perceived loudness vs bandwidth of broadband noise at the same power: if the bandwidth is larger than the critical frequency, its perceived loudness increases with the bandwidth.

bandwidth $\Delta f = f_2 - f_1$ [see Appendix B(3)]. The band-intensity level (L_{band}) is the integrated sound power over all frequencies within the band [see Eq. (3.7)]

$$L_{\text{spectral}} \equiv 10 \log \frac{I_1}{I_0} \doteq L_{\text{band}} \equiv 10 \log \frac{I_2}{I_0} + 10 \log \frac{\Delta f}{1\text{Hz}}, \qquad (3.7)$$

where $I_0 = 10^{-12}$ W/m^2. The band intensity level and the spectral level are equal at a given center frequency when Eq. (3.7) is satisfied. Two bands have the same sound spectral "intensity" when their band intensity levels are equal.

★ Example 3.19: Human ears perceive pitch in logarithmic scale. The basic unit for the music scale is the "octave" for a frequency ratio of 2. The octave band corresponds to a bandwidth of $f_1 = 2^{-1/2} f_c$ and $f_2 = 2^{1/2} f_c$. What is the bandwidth of the 1/3 octave band?

Answer: The cutoff frequencies of the 1/3 octave band are $f_1 = 2^{-1/6} f_c$ and $f_2 = 2^{1/6} f_c$ with $f_2/f_1 = 2^{1/3}$. The bandwidth is $\Delta f = f_2 - f_1 = 0.2316 f_c$.

Figure 3.6(b) schematically shows the loudness level vs bandwidth of the broadband noise intensity level at equal sound spectral intensity. If the noise bandwidth Δf exceeds the critical bandwidth, the perceived loudness will continue to increase with Δf. The result can be summarized as follows: (1) A band of complex tone within a critical bandwidth will have the same perceived loudness as that of a single tone at 1 Hz width if they have the same sound spectral "intensity." (2) When the

frequency width of a complex tone, at the same spectral *intensity*, is
more than the critical bandwidth, it will sound louder than a pure tone
that has the same *spectral intensity* of Eq. (3.7). These psychoacoustic
experiments were carried out by Fletcher *et al.* in 1940s. Fletcher's
results suggested a method for estimating the width of critical bands.

The critical bandwidth (CBW), found by psychoacoustic experiments, varies from 100 Hz at low frequency to 3,500 Hz at 15 kHz:

$$\text{CBW [Hz]} \approx 94 + 71 \left(\frac{f_c \text{ [Hz]}}{1000 \text{ [Hz]}} \right)^{1.5}. \tag{3.8}$$

The CBW is generally independent of intensity and depends only on the
center frequency. It is nearly constant at low frequency, lying between
the whole tone and the $\frac{1}{3}$ octave at high frequency, shown in Fig. 3.7(a).
It is said that 30 such bands will span the human audible range.[10]

According to the Greenwood characteristic frequency (CF) formula,
two pure tones spaced one octave apart, e.g. C5 and C6, have little
overlap on the cochlear membrane. Figure 3.7(b) shows the distance on
the cochlear membrane from the apex vs the CF of an octave assuming
a cochlear length of 35 mm. Since the CFs of C5 and C6 separate
well on the cochlear membrane, few hair cells respond to both frequencies simultaneously; there is little interference in processing these two
signals. When the frequencies of two pure tones approach each other,
the distance between the hair cells responding to the two frequencies
becomes smaller. When two pure tones are close in frequency or lie
within the critical band, there is considerably overlap in their amplitude envelope on the basilar membrane. The critical bandwidth (CBW)
is "the minimum frequency distance for which two tones can excite the
same hair cells." When the frequency difference between two notes falls
within the CBW, these two notes may excite the same hair cell on the
cochlear membrane.

The cochlea is a superposition of filter bands. When two frequencies
fall on the same filter band, they can mask each other. Using the notch-
filter on the tuning curve (Fig. 3.3) of the cochlea to explore the sound

---

[10] A comprehensive review is available by B. Scharf, Critical Bands, pp. 159–202,
in *Foundations of Modern Auditory Theory*, edited by Jerry Tobias (Academic
Press, NY, 1970).

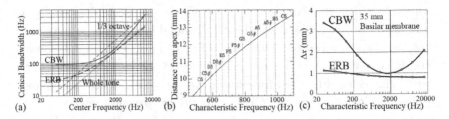

Figure 3.7: (a) The critical bandwidth (solid line) vs the frequency, compared with the musical whole tone (dotted line) and 1/3 of the octave band (short dashes). The long dashed line shows the equivalent rectangular band (ERB). (b) The distance from the apex in [mm] for a cochlear length of 35 mm vs the 5th octave characteristic frequency according to Greenwood's formula. (c) The width of the CBW and ERB vs CF for a basilar membrane of 35 mm.

masking, one obtains the equivalent rectangular band (ERB), given by

$$\text{ERB} = 24.673 \times (0.004368f + 1) \text{ [Hz]}, \qquad (3.9)$$

where f is the center frequency in [Hz]. Using the Greenwood CF formula, we can convert the CBW and ERB into the width on the basilar membrane. Figure 3.7(c) shows the width on the cochlear membrane Δx for CBW and ERB vs CF. We find the CBW varies from 3 mm to 1 mm on the basilar membrane, while the width of the ERB is nearly constant at about 0.9 mm. There are 40 ERBs for the total length of the cochlear membrane in our hearing range of 20–20,000 Hz. At high frequency $f > 500$ Hz, we find ERB $\approx 0.108f$, which is slightly smaller than the whole tone at $\Delta f = 0.122f$.

★ Example 3.20: Use the Greenwood CF formula in Eq. (3.1) to find the length of cochlea that covers the ERB at $f_c = 500$ Hz, assuming the length of cochlea is 35 mm.

Answer: From Eq. (3.9), we find $\Delta f = \text{ERB} = 78.6$ Hz at $f_c = 500$ Hz. Using Eq. (A.9), the upper and lower frequencies of the band are 541 and 462 Hz, respectively. The upper and lower positions of the characteristic frequencies are $541 = 165.4 (2^{\alpha x_2} - 0.88)$ and $462 = 165.4 (2^{\alpha x_1} - 0.88)$, from Eq. (3.1) of the Greenwood CF formula. Solving these equations, we find $x_2 - x_1 \approx 0.025 = 1/40$, i.e. there are 40 ERBs in the length of the cochlea. If the length of the cochlea is 35 mm, this band length is 0.87 mm, shown in Fig. 3.7(c).

When the frequencies of uncorrelated tones are within the critical band, the resulting intensity is $I = I_1 + I_2 + I_3 + \cdots$, where $I_1, I_2, \ldots$ are the intensities of individual tones. The loudness level of the combined level tone is determined from the sum of the intensities. However, when we listen to two tones having the same sound pressure level, the loudness of the combined tones will begin to increase if the frequency difference is larger than the critical bandwidth. Broadband sounds, such as jet aircraft, seem louder than pure tones or narrow band noise having the same sound pressure level. As the bandwidth of the combined sound exceeds the critical bandwidth, the resulting loudness may be higher than that obtained from the summation of the intensities. As the bandwidth increases further, the loudness may approach $S = \sum S_i$, the sum of the sones.

Many psychoacoustic experiments have established that our ear averages sound energy over about 200-ms time intervals. The sound loudness increases with increasing duration up to about 200 ms. The loudness level increases by 10 dB when the time duration increases from 20 ms to 200 ms (i.e., by a factor of 10). The sound level saturates at about 200 ms.

VI.   Masking

Masking is a psychoacoustic phenomenon that when two or more tones are simultaneously heard together, a lower frequency tone can mask higher frequency tones. The masking arises from the experimental facts that the tuning curve of each individual neuron is sharply cut off at the low frequency end [see Fig. 3.2(b) and 3.3(b)]. These neuron tuning curves show that high-frequency tones do not reach far into the characteristic frequency of the cochlear membrane, and thus the low-frequency sounds will mask high frequency tones. Figure 3.8(a) shows the schematic responses of a low-frequency pure tone B and a high-frequency pure tone A on the basilar membrane. These four cases are respectively: no masking when two tones do not overlap, tone B masks tone A, tone B completely masks tone A, and finally the more intense tone A is not fully masked by tone B.

Figure 3.8(b) shows the masking of a 1-kHz pure tone by white noise at various SPLs. At low-noise (quiet) conditions, the perceived loudness

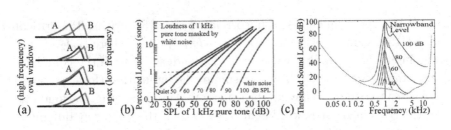

Figure 3.8: (a) Schematic drawing of tuning curve of cochlear membrane showing that the low frequency tone (B) can mask high frequency tone (A). (b) A perceived loudness (in sone) under various white noise level at 50–100 dB SPL. (c) The threshold of hearing under the masking of 1 kHz at various SPL. Note that the masking is more significant at high frequency.

is 1 sone at 40 dB, 2 sone at 50 dB, 4 sone at 60 dB, etc. (see Table 3.3). At 80 dB white noise, it requires 68 dB of 1-kHz tone to be perceived as 1 sone. At 100 dB white noise, 86 dB of pure tone is needed to be perceived as 1 sone. Figure 3.8(c) shows the SPL of the threshold of hearing masked by a 1-kHz pure tone at various masking SPLs. Note that the 1-kHz pure tone does not efficiently mask low-frequency tones, but it severely masks tones for frequencies higher than 1 kHz.

VII.   Binaural Hearing, Localization and Haas Effect

Localization is the ability that we can determine the direction of a sound source. This ability rests on binaural hearing that we hear with both ears located on opposite sides of our head. The localization is the result of the responses of auditory neurons to "inter-aural time difference (ITD)" and "inter-aural intensity differences (IID)." Figure 3.9(a) shows shadow effects at high frequency and diffraction at low frequency. The shadow effect is important at high frequency $f > v/a$, where v is the speed of sound and a is the width of obstruction [see Eq. (2.10)].

In particular, the pinna casts shadow for high frequency tones $f > v/a \sim 5{,}000$ Hz, where a is the size of the pinna. This intensity difference aids to differentiate sounds coming from the front or back. On the other hand, diffraction becomes important at low frequencies $f < v/a \sim 2{,}000$ Hz, where a is the size of the human head. Time-delay

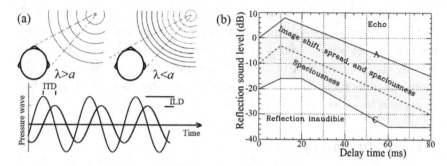

Figure 3.9: (a) (Top) A schematic drawing of binaural hearing showing the diffraction effect at low frequency ($\lambda > a$) and shadow effect at high frequency ($\lambda < a$). (Bottom) Localization of a sound source results from Inter-aural Time Difference (ITD) and the Inter-aural Level (or Intensity) Difference (ILD or IID) from both ears. (b) Effects of the reflection level vs the delay time. Recording engineers employ the Haas effect to enrich our perception of a broader and more spacious sound from a recording by having the reflected sound level and delay time be in the region bounded by curves A and C.

experiments have confirmed that localization relies on the detection of phase difference between the two ears at frequencies $f < 1,000$ Hz.

Binaural psychoacoustic experiments also show the "precedence effect," or *the law of the first wave front*, known as the *Haas effect*. In 1949, Helmut Haas presented his doctoral dissertation at the University of Gottingen Germany on systematic experiments about the binaural psychoacoustic precedence effect. [Dr. Ing. K.P.R. Ehrenberg translated his thesis into English in 1949, and his translation was reproduced in the United States as "The Influence of a Single Echo on the Audibility of Speech," *J. Audio Eng. Soc.*, Vol. **20**, 145 (1972)]. When listeners hear a sound and its identical copy, the perceived spatial location is the first-arriving sound. This precedence effect is true if (1) the levels of the sound and its copy differ by no greater than 10 dB, and (2) the time delay between them is below the echo threshold of about 35 ms (sometimes 2–50 ms, even up to 100 ms for musical sound). Although the time-lagging sound may also affect the perceived location, its effect is suppressed by the first-arriving sound. Figure 3.9(b) shows the schematic drawing of the Haas effect. When the reflection sound level and the time delay are within a certain range, the perceived sound image becomes spacious. Recording engineers use the Haas effect to produce sound that tricks our brain into perceiving a sound with width and space. If the minor of two identical second sources has an intensity level within about 10 dB and is delayed by no more than 50 ms, the combination of these two sounds is perceived as a single, more spacious and richer sound, as schematically shown in Fig. 3.9(b) in regions bounded by curves A and C.

VIII. Summary

Three subjective attributes of sound are loudness, pitch, and timbre. Their corresponding physical parameters are pressure, frequency, and spectrum and pulse envelope. Pulse duration is also important in pitch and loudness. The physical parameter most directly connected with loudness is sound pressure (in pascals) and sound intensity (in W/m^2). The human auditory system can detect sound pressure disturbance at 20 μPa. Damage to our hearing system may occur when the sound pressure is higher than 20 Pa. Our ears are sensitive to frequencies ranging from 20 Hz to 20 kHz.

We divide the human ear into the outer, middle and inner ears. The outer ear resembles a closed cylindrical tube that boosts the sensitivity of hearing frequencies around 1–4 kHz. The middle ear consists of the system of ossicular bones (malleus, incus and stapes) that converts the wave energy into mechanical vibration. It provides sound amplitude amplification by a factor of about 17 (or 25 dB). The middle ear also has built-in acoustic reflexes to avoid injury by very loud sounds. The stapedius muscle stiffens the ossicular chain by pulling the stapes (stirrup) of the middle ear away from the oval window. The tensor tympani muscle stiffens the ossicular chain by pulling the malleus (hammer) toward the middle ear and loading the eardrum. This will dampen the oscillation amplitude of the tympanic membrane. However, the reflex time is about 30–40 ms, and full protection time is about 150 ms, which is not effective to protect against a loud impulse. A sound of 90–100 dB, preceding a loud pulse, can arm the acoustic reflex and protect our ears. We also activate the stapedius reflex during our speech. The activation of the stapedius reflex reduces sound intensities reaching the inner ear by about 20 dB, minimizing the interference caused by hearing our own voice.

The inner ear is composed of semi-circular canals and the cochlea. The cochlea converts mechanical vibration to an electric neural signal transmitted through the chain of structures including the cochlear nucleus, superior olive, lateral lemniscus, inferior colliculus, and medial geniculate body to the auditory cortex in the brain on both sides of auditory neural network. The Greenwood characteristic frequency of the tuning curve depends on the position of the cochlear membrane, supporting the "place theory of pitch," while the electric neural signals retain the time structure of the sound signal, as evidence for the "temporal theory of pitch."

Because our hearing system has a very large dynamic range, we use the logarithmic decibel (dB) scale for sound levels given by Eqs. (3.3) to (3.6). The unit for power level, sound pressure level, and sound intensity level is the decibel (dB). The sound intensity level and sound pressure level are equal. We simply call them "sound level." The loudness level (phon) is equal to the sound level at 1 kHz on the equal loudness curve; meanwhile, the loudness (sone) provides the psychoacoustic loudness to human ears. The ISO-226 equal loudness curve (or Fletcher–Munson curve) set the standard for loudness measurements, classified into dBA, dBB, dBC, ... filtered scales. The sensation of hearing (psychoacoustics) obeys both the Fechner's logarithmic law and Stevens's power law. The sensation is still a complex function of frequency. The cochlea is a superposition of filter bands. When two frequencies fall on the same filter band, they can mask each other. Psychoacoustic experiments were used to determine the critical bandwidth (CBW) and the equivalent rectangular band (ERB), given by Eqs. (3.8) and (3.9). When we hear two or more tones, our common experience indicates that low frequency tone will mask the higher frequency ones. The sharp cutoff at the low frequency end of the tuning curve can explain this phenomenon.

Binaural hearing is essential to sound localization. Our ability to determine the location of a sound source depends on the binaural hearing with different mechanisms of localization for low and high frequencies. At low frequencies, the phase (time) difference plays an important role in identifying the sound direction. At high frequencies, the inter-aural intensity difference (IID) plays an important role in identifying the source direction. For low frequency sound, we employ inter-aural time difference (ITD) to identify the location of sound sources. An interesting binaural psychoacoustic effect is the precedence effect or law of the first wave front, known as the Haas effect. Recording engineers employ the Haas effect to provide the perception of spacious sound.

IX. Homework 03

1. The characteristic frequency at the mid-point of human cochlear membrane is _____ Hz.

2. The function of the middle ear is to amplify the pressure delivered to the oval window, and to *protect* the inner ear through the

   _____ _____ .

3. The _____ scale is commonly used for comparing sound power level, sound intensity level, and sound pressure level.

4. Find the following logarithms: $10 \times \log 2 =$ _____, $10 \times \log 4 =$ _____, $10 \times \log 200 =$ _____ .

5. The _____ of a "number" is the power to which 10 (the base) must be raised to give the "number."

6. Given $\log x$, find the number x: $\log x = 0.3$, $x =$ _____; $\log x = 3.00$, $x =$ _____ .

7. The sound intensity level is directly connected physical parameter of sound wave _____ .

8. Loudness level (or equal loudness level) depends on the sound wave _____ and on the _____ of the sound wave.

9. Two sound sources produce sound power levels of 53 dB and 66 dB relative to the reference power level $W_0 = 10^{-12}$ W, Find $W_1/W_0 =$ _____, $W_2/W_0 =$ _____, and $W_2/W_1 =$ _____ .

10. A sound of intensity $I = 10^{-6}$ W/m^2 falls on a detector of area $A = 7 \times 10^{-5}$ m^2. (This is about the size of an eardrum.) (a) The

sound level is _____ dB. (b) What is the power on the eardrum? (c) What is the rms pressure on the eardrum? (d) If the sound continues for 30 s, the amount of energy received is _____ J.

11. Given a sound with $L_p = 50$ dB, the sound pressure is _____ N/m^2 and the sound intensity is _____ W/m^2.

12. Two sounds differ in level by 46 dB. The ratio of their intensities is _____ .

13. The pressure of a sound wave in air can vary from $-\Delta p_{max}$ to Δp_{max}, where Δp_{max} is the amplitude of the sound pressure. If the maximum amplitude of sound pressure is $\Delta p_{max} = p_{atm}$, the atmospheric pressure, it will cause cavitation of air. This means that there is no sound wave with pressure amplitude higher than the atmospheric pressure. The sound pressure level at this condition is _____ dB.

14. At a rock concert, a dB-meter registers 125 dB when placed 3.0 m in front of a loudspeaker on the stage. The intensity of the sound at 3 m is _____ W/m^2. The sound power output of the speaker is _____ W. The sound intensity at 30 m away from the speaker is _____ dB. The distance from the speaker at which the sound level reduces to 99 dB is _____ m. [Assume uniform spherical spreading of the sound from the speaker.]

15. For a sound wave at the pain threshold ($L_p = 130$ dB), the sound pressure = _____ N/m^2 and the force on an eardrum with a diameter of 8.0 mm = _____ N.

16. For a speaker whose sound power output is 140 dB. Find the distance from the speaker that will reduce the sound level to 99 dB. _____ m.

17. The trombone bell has an area of 0.10 m^2. If the power radiated from the bell during a very loud note is 1.5 W, the sound intensity is _____ W/m^2 and the sound intensity level at the trombone bell is _____ dB.

18. If the whirring of a single fan produces a sound intensity level (SIL) of 50 dB, how many identical fans operate simultaneously to make a SIL of 63 dB? $N =$ _____ .

19. With one violin playing, the sound level at a certain place is 50 dB. With 2 violins playing, the sound level = _____ dB and, with 4 violins playing, the sound level = _____ dB.

20. A loudspeaker uses 10.0 W of electrical power. It has an efficiency of 1.0% for converting this electrical power to sound power. The sound power output of the speaker = _____ W. This corresponds to a sound power level = _____ dB. At a distance of 2.0 m, the sound intensity is _____ W/m^2, the sound intensity level is _____ dB, and the rms sound pressure is _____ N/m^2. The sound intensity at a distance of 4.0 m from the speaker is _____ W/m^2. (We assume that the sound from the speaker radiates equally in all directions.)

21. What is the loudness level of a pure tone at 400 Hz 70 dB SPL? _____ phon.

22. What is the loudness of a pure tone at 400 Hz 70 dB SPL? _____ sone.

23. Two pure tones that are close enough in frequency to overlap in their response on the basilar membrane are said to lie within the same _____ band.

24. The smallest change in sound level (intensity limen) detectable by a person with normal hearing is 1 dB. The percentage difference in sound intensity is _____ %.

25. You are attending a rock concert wearing earplugs that provide a 13 dB reduction in sound level. The percentage of the sound energy getting through the plugs is _____ %.

26. The middle ear is composed of three bones, malleus or hammer, incus or anvil, and stapes or stirrup. It provides amplification of the transmitted pressure by a factor of 17, this corresponds to amplification of _____ dB.

27. The critical bandwidth (CBW) of Eq. (3.8) at $f_c = 1$ kHz is 165 Hz. What are the lower and higher band frequencies: $f_1 =$ _____ Hz and $f_2 =$ _____ Hz.

28. In the Example 3.20, the ERB is 78.6 Hz at $f_c = 500$ Hz. Show that the lower and high band frequencies are 462.2 Hz and 540.8 Hz respectively.

29. Much of our ability to determine the direction of a sound depends on _____ hearing, the fact that we have two ears located on opposite sides of our head.

30. Localization for _____ frequency sound-source relies on Inter-aural Time Difference (ITD) and for _____ frequency sound-source relies on the Inter-aural Level Difference (ILD).

31. When combining independent, uncorrelated sound sources, we add sound _____ to obtain the combined excitation of the air.

32. A binaural psychoacoustic effect is the precedence effect or law of the first wave front, known as _____ effect. Recording engineers employ this effect to provide a more spacious sound for our hearing perception.

33. The acoustic reflex is the action of muscles (tensor tympanic muscle and stapedius muscle) tighten to provide protection of the inner ear. This process happens in the _____ ear.

34. When we hear two or more tones, the _____ frequency tone is more likely to mask the _____ frequency tone.

35. High frequency sound excite the basilar membrane _____ to the oval window.

36. The loudness S in [sone] is a power function of pressure by $S = Cp^{0.6}$ at 1 kHz pure tone, where $C = 42$ [sone], and the rms sound pressure p is in [Pa]. At a sound level 40 dB, the rms sound pressure is _____ [Pa] and the loudness is _____ [sone]. At sound level of 60 dB, the rms sound pressure is _____ [Pa] and the loudness is _____ [sone].

Chapter 4

Pitch, Timbre and Musical Scale

Besides loudness, other attributes of sound to human sensation are pitch and timbre. This chapter discusses pitch, timbre, Fourier theorem and frequency spectrum, combined tones, and harmony. Pitch is "the attribute of auditory sensation in terms of which sounds may be ordered on a scale extending from low to high"; and timbre or tone quality is "that attribute of auditory sensation in terms of which listeners can judge two sounds similarly presented and having the same loudness and pitch as dissimilar." — The American National Standards Institute.

I. Pitch

Human ears can detect sound in the frequency range from 20 to 20,000 Hz, sense sound levels down to 0 dB, and tolerate sound levels up to 120 dB. The most sensitive frequency to our ears is in the range of 1,000 to 4,000 Hz (see ISO-226 in Fig. 3.4). All cultures in human societies create language and music with specific sound patterns for communication, art, and entertainment.

The octave unit, at a frequency ratio of 2:1, is the basis of the music scale in all cultures. The European diatonic music scale is based on 7 notes and a repeated octave, i.e. C*D*EF*G*A*BC for a complete cycle. To play music together, a pitch standard set the 4th octave A-note ($A4$ or A_4) at 440 Hz. Historically, the A4 note varied from 400 to 458 Hz and gradually converged to 440 Hz. In 1936, the American Standards Association recommended 440 Hz for the A4 note. In 1939, an international conference in London adopted the standard. The ISO took up this standard in 1955 in the publication ISO-16 (reaffirmed in 1975). Although this is not a universal standard (e.g. some European musicians and orchestra prefer 442 Hz or 446 Hz for A4, while some prefer a lower frequency), musicians and musical instrument makers almost universally adopt this pitch standard. The US Bureau of Standards broadcasts 440 Hz pure tone on its shortwave radio station WWV.

A.   Absolute Pitch (AP) or Perfect Pitch

Absolute pitch is the ability to identify a given musical pitch without a reference tone. Less than 0.01% of the general population have this special ability to recognize absolute pitch, while 98% of the population have absolute color recognition. Four theories on absolute pitch are (1) heredity, (2) learning through training and practice, (3) unlearning by emphasis on relative pitch, and (4) imprinting by learning at young age. Those with AP tend to start their musical education in early childhood, at age 6 or younger, and it is nearly impossible to acquire absolute pitch as an adult. Survey data show that possession of absolute pitch is not a prerequisite for outstanding musicianship.[1] In contrast to absolute pitch, there is an indication that 2–5% of general population have problems in pitch perception.[2]

B.   Pitch Scales

The basic unit in most musical scales is the octave at frequency ratio 2:1. One can divide the octave into different pitch units. The diatonic scale is composed of 7 pitches and a repetitive octave. The *equal temperament tuning* subdivides the octave into 12 intervals called semitones, with names A through G, including sharp and flat. A whole tone is equal to two semitones (see Fig. 4.1). Each semitone is divided into 100 cents, where 1 cent corresponds to an increment of frequency by

---

[1] Some AP self-test websites are http://www.audiocheck.net/blindtests_abspitch. php; http://deutsch.ucsd.edu/psychology/pages.php?i=6215. See also E.A. Athos, B. Levinson, A. Kistler, J. Zemansky, A. Bostrom, N. Freimer, and J. Gitschier, *PNAS*, **104**, 14795 (2007).

[2] A self-test of pitch perception website at NIDCD is available at https://www. nidcd.nih.gov/tunestest/test-your-sense-pitch.

multiplying $2^{(1/1200)} = 1.000578$. Each higher semitone is obtained by multiplying $2^{(100/1200)} = 1.0594$ to the frequency, and one whole tone by $2^{(200/1200)} = 1.12246$.

Figure 4.1: A diatonic scale is an eight-note musical scale composed of seven pitches and a repeated octave. The diatonic scale includes 5 whole-steps and 2 half-steps for each octave. The two half steps, i.e. EF and BC, are maximally separated from each other in the chromatic scale C*D*EF*G*A*BC, where asterisks represent the sharp or flat note of the adjacent notes. The right plot shows the conventional music staff notation.

The frequency of most wind instruments change with temperature and humidity. The velocity of sound in air changes 0.6 m/s for each Celsius degree. Accordingly, the pitch rises "3 cents" per °C temperature rise. Conversely, string instruments tend to lower its pitch as temperature rises.

★ Example 4.1: Verify that a 1°C temperature change in wind instruments is a scale change by 3-cents.

Answer: The frequency of a wind instrument is proportional to the speed of sound. If the musical instrument is tuned to perfect pitch f at 20°C, then a 1°C increase will cause the speed of sound to increase from 343 m/s to 343.6 m/s. The frequency increment is $343.6/343 = 1.00175$, which is nearly equal to $2^{3/1200} = 1.00173$ for 3 cents of change in music scale.

★ Example 4.2: The human ear covers the frequency range of 10 octaves. What does it mean?

Answer: Our lowest and highest perceivable frequencies are 20 and 20,000 Hz respectively. The 10th octave above 20 Hz is $20 \times 2^{10} = 20,480$ Hz, i.e. 10 octaves cover our hearing range.

★ Example 4.3: List the standard center frequencies of the noise octave bands for acoustic experiments and find the corresponding frequency range of each octave band [see Appendix B(4)].

Answer: The center frequencies for the 10 octaves in standard acoustic experiments are normally set at

f_c (Hz)	31.5	63	125	250	500	1,000	2,000	4,000	8,000	16,000

An octave band has frequency range f_1 and f_2, such that $f_2/f_1 = 2$ and $f_1 f_2 = f_c^2$. Thus, the audible octave bands are $(f_1 f_2) = (2^{-1/2} \times 31.5,\ 2^{1/2} \times 31.5)$, $(2^{-1/2} \times 63,\ 2^{1/2} \times 63)$, $(2^{-1/2} \times 125,\ 2^{1/2} \times 125)$, $(2^{-1/2} \times 250,\ 2^{1/2} \times 250)$, $(2^{-1/2} \times 500,\ 2^{1/2} \times 500)$, $(2^{-1/2} \times 1,000,\ 2^{1/2} \times 1,000)$, $(2^{-1/2} \times 2,000,\ 2^{1/2} \times 2,000)$, $(2^{-1/2} \times 4,000,\ 2^{1/2} \times 4,000)$, $(2^{-1/2} \times 8,000,\ 2^{1/2} \times 8,000)$, and $(2^{-1/2} \times 16,000,\ 2^{1/2} \times 16,000)$, respectively. Each band has a bandwidth of 1 octave. This covers the entire audible range from 20 to 20,000 Hz. To provide finer resolution in frequency, one can use $\frac{1}{2}\ \frac{1}{3}$, $\frac{1}{4}$, $\frac{1}{6}$ ... octave bands. For example, if one wishes to study finer frequency resolution at 500 Hz by using $\frac{1}{4}$ octave bands, the frequency ranges are $(2^{-1/2} \times 500,\ 2^{-1/4} \times 500)$, $(2^{-1/4} \times 500,\ 500)$, $(500,\ 2^{1/4} \times 500)$, and $(2^{1/4} \times 500,\ 2^{1/2} \times 500)$. Each band has $\frac{1}{4}$-octave bandwidth. The total bandwidth of these four $\frac{1}{4}$-octave bands is an octave.

C.   Psychoacoustic Pitches: MEL and Bark

In some psychoacoustic experiments, an average listener listened to a pure tone of 4,000 Hz and then listened to a low frequency tone. The listener was then asked to choose a pitch that was halfway between 0 and 4,000 Hz. The listener would choose 1,000 Hz, i.e. 1,000 Hz was judged to be halfway between 0 and 4,000 Hz. This subjective pitch unit is called the melody scale (mel), set at 1,000 Hz = 1,000 mel. It is fitted by Ward's formula $m = 2,595 \log_{10}(1 + f/700)$, devised by W. Dixon Ward on "Musical Perception," shown in Fig. 4.2(a).

Another psychoacoustic unit is bark, defined as the critical bandwidth, i.e. 1 bark = 100 mels. It is argued that the frequency range of hearing can be subdivided into 24 bands with bandwidth Δf_k, where $\Delta f_k \sim 1$ bark. Using the Greenwood CF formula for the corresponding mel scale, we plot one bark width on the basilar membrane vs frequency in mel in Fig. 4.2(b). The width varies from 0.7–2.1 mm for a basilar membrane of 35 mm. One bark is about 1/20 of the basilar membrane at low frequency and about 1/50 of the basilar membrane at high frequency. Both the mel and bark scales are not uniformly distributed on the basilar membrane. We can compare the width of the bark scale with that of CBW and ERB in Fig. 3.7(c).

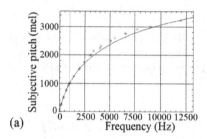 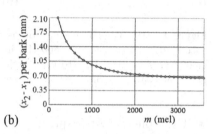

Figure 4.2: (a) Mel vs frequency in linear scale, where psychoacoustic survey data are compared with the Ward's formula. (b) Width of 100 mels on the basilar membrane vs the mel scale using Greenwood CF formula with basilar membrane length of 35 mm.

D. Pitch Discrimination, Pitch Limen, Just Noticeable Difference (jnd)

Psychoacoustic experiments show that the pitch limen or the just noticeable difference (jnd) of a pure tone is about 3.6 Hz below 500 Hz. The pitch limen increases to about $0.007f$ for a pure tone at high frequencies. Figure 4.3 shows (a) the pitch limen with the critical bandwidth and (b) the fractional $\Delta f_{\text{jnd}}/f$ vs frequency for various sound levels. Note that $\Delta f_{\text{CBW}} \approx (20 \sim 50)\Delta f_{\text{jnd}}$. Since the critical bandwidth varies from $1 \sim 3$ mm on basilar membrane, shown in Fig. 3.7(c), the limen of the pitch has a length of about 50 μm on the basilar membrane (see Fig. 4.3 and the Example 4.4). Human ear can hear 10 octaves with about 700–1000 jnds in 35 mm length of the basilar membrane.

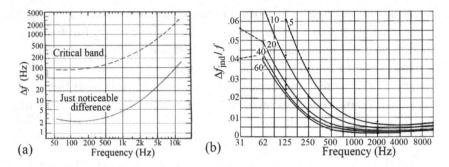

Figure 4.3: (a) The survey data of critical bandwidth and just noticeable difference (jnd) vs frequency. At 500 Hz, jnd is about 3.6 Hz, while the jnd is about
0.007f at $f > 1,000$ Hz. (b) the fractional jnd vs frequency at various SPL from
psychoacoustic experiments.[3]

★ Example 4.4: At frequency $f > 1$ kHz, the jnd is about $0.007f$. Use
the Greenwood CF formula of Eq. (3.1) to find out the distance
between hair cells that resolve this frequency limen, assuming the
cochlear length is 35 mm.

Answer: Since $\Delta f/f = 0.007$, we find $f_2/f_1 = 1.007$. Using the
Greenwood CF formula at high frequency, we obtain $2^{\alpha\Delta x} = 1.007$,
or $\alpha\Delta x = \mathrm{Log}(1.007)/\mathrm{Log}(2) = 0.010$. Using $\alpha = 7.0$, we find that
the distance between hair cells for jnd frequencies is $d = (35 \text{ mm}) \times
\Delta x \approx 50 \ \mu\text{m}$, which is about a few hair cells in width. The jnd of
the low-frequency region in Fig. 4.3(a) also gives $d \approx 40 \ \mu\text{m}$.

E.   Pitch of Pure Tone vs Sound Intensity Level

The pitch of a pure tone may depend on its intensity; psychoacoustic
experiments show that, higher intensity is perceived as higher pitch
at frequencies $f > 2$ kHz, while the effect is reversed at frequencies
$f < 2$ kHz (see Fig. 4.4). There is little dependence of pitch vs intensity
at $f = 2$ kHz. Note that 2 kHz tone happens to excite the mid-point of
the cochlear membrane [see Fig. 3.3(a)]. Furthermore, we perceive little
pitch change vs intensity for a sound with complex waveform.

---

[3]See E.G. Shower and R. Biddulph, Differential pitch sensitivity of the ear, *JASA*,
3, 275 (1930).

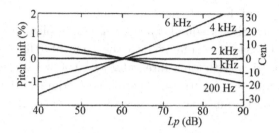

Figure 4.4: Our ears seems to recognize the 2 kHz pitch independent of the sound loudness level. A frequency above 2 kHz, a louder sound appears to have higher pitch, while at frequencies below 2 kHz; a louder sound appears to have lower pitch.

II. Fourier Theorem, Uncertainty Principle and Perception of Pitch

In 1822, Joseph Fourier developed a frequency analysis method to analyze any function of time and space, now called Fourier analysis to his honor [see Appendix B(2)]. A corollary of Fourier analysis is a Fourier theorem stating that "Any periodic waveform with period T_0 can be decomposed into sinusoidal waves with appropriate amplitude and phase at frequencies that are integer harmonics of the fundamental frequency $f_0 = 1/T_0$."

★ Example 4.5: A complex wave repeats every 8 ms. What is the fundamental frequency? What are the frequencies of the 2nd and 3rd harmonics of the wave?

Answer: Since the period is $T_0 = 8$ ms, the fundamental frequency is $f_0 = 1/(0.008\,\text{s}) = 125$ Hz. The frequencies of the 2nd and 3rd harmonics are $2f_0 = 250$ Hz and $3f_0 = 375$ Hz, respectively.

★ Example 4.6: In speech, the vocal folds close and open so that air pulses have a fixed regular period. The pitch of human speech is about 100, 200, and 300 Hz for men, women, and children, respectively. What are the corresponding periods of air pulses in speech?

Answer: The waveform of the speech is triangular, and the periods of the waveforms are $1/(100\ \text{Hz}) = 10$ ms, $1/(200\ \text{Hz}) = 5$ ms, and $1/(300\ \text{Hz}) = 3.33$ ms for men, women and children, respectively. The period of vibration is the resonance of our vocal folds.

An example of Fourier theorem is shown in Fig. 4.5, where the left-bottom plot shows sound wave of a violin playing C4 note for a duration of 0.844 s. Zooming into the time at 0.63 s, we observe the wave repeating with time interval $T_0 = 0.00382$ s. The right plot shows the frequency spectrum, where the fundamental frequency is $f_0 = 1/T_0 = 262$ Hz and its harmonics are nf_0, $n = 1, 2, 3, \ldots$.

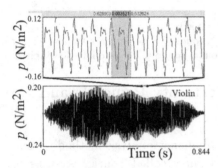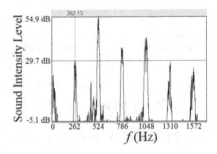

Figure 4.5: The waveform (left) and spectrum (right) of a violin C4 note, obtained from the Praat program [Ref. 2]. The top-left plot shows the detailed waveform at 0.63 s that repeats with period $T_0 = 0.00382$ s. The corresponding fundamental frequency is 262 Hz.

A. Pitch Recognition vs Time Duration and Uncertainty Principle

Early psychoacoustic experiments using the Félix Savart wheel indicated that one could perceive successive clicks in just 2 ms. Subsequent experiments showed that a longer duration was required. Figure 4.6(a) shows data obtained by Bürck, Kotowski, and Lichte (BKL),[4] where one needs 12–60 ms of time duration to resolve a pitch. The minimum time for our ears to recognize a pitch is longer than 10 ms. Translating the BKL data into number of oscillation periods, the perceived pitch is more than 4 periods at frequencies below 200 Hz and increases to 250 periods at 10 kHz.

---

[4]W. Bürck, P. Kotowski and H. Lichte, *Elect. Nachr. Techn.*, **12**, 278 (1935).

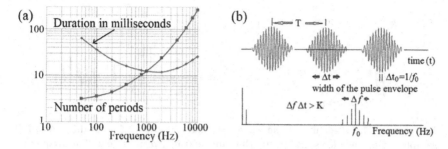

Figure 4.6: (a) Psychoacoustic data of Bürck, Kotowski, and Lichte (BKL) on time duration for pitch recognition in milliseconds and the corresponding number of periods vs frequency. (b) Schematic drawing of periodic sinusoidal pulses at fundamental frequency f_0, separated by the time T. The envelope of the pulse has a width Δt. The lower plot shows its Fourier spectrum.

The perceived pitch of a pure tone may also depend on the pulse envelope, loudness, and the interfering tone or noise. Figure 4.6(b) shows a schematic drawing of periodic sinusoidal pulses at frequency f_0 with an envelope pulse length or "FFT time window" of Δt. The lower plot shows its corresponding Fourier spectrum, where the "frequency spread" Δf, centered around f_0, is related to the pulse width Δt via the uncertainty principle:

$$\Delta f \Delta t \geq K = \frac{1}{4\pi} \sim 0.080\,. \tag{4.1}$$

To achieve a frequency resolution of $\Delta f = 1$ Hz, the FFT time window must be longer than $\Delta t \geq \frac{1}{4\pi(\Delta f)} \approx 80$ ms. Figure 4.7 demonstrates the uncertainty principle by analyzing the waveform of the violin C4 shown in Fig. 4.5. Figure 4.7 shows the FFT frequency spectra vs frequency at time window $\Delta t = 20$, 100 and 500 ms, respectively. Note that the frequency resolution improves as the FFT sampling time window increases until it reaches the intrinsic limit of the sound source.

The best that our ears can resolve is the "jnd." The jnd psychoacoustic experiments are based on measurements with very long pulses of sound and concern the recognition of two frequencies being different. "Pitch recognition" is the ability of the human ear to recognize a pitch. From the BKL data in Fig. 4.6(a), it requires 12 ms duration to perceive a pitch at $f \sim 2{,}000$ Hz. This corresponds to a frequency resolution of 6.6 Hz from the uncertainty principle.

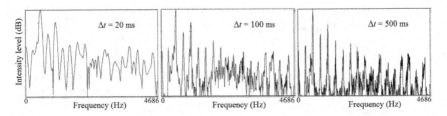

Figure 4.7: The spectrum of the waveform shown in Fig. 4.5 with different FFT sampling time windows Δt = 20, 100, and 500 ms, respectively. The frequency resolution improves when the sampling time increases until it reaches its intrinsic resonance width.

B.   Pitch of Complex Tones (Virtual or Residual Pitch)

The pitch that we perceive from a complex tone, made of several pure tones at frequency integer multiples of each other, may differ from the frequency content of the complex tone. We call the perceived pitch the "virtual pitch" or "residual pitch." For example, when two pure tones with frequencies 200 Hz and 300 Hz are played together, as shown in Fig. 4.8, our ears perceive the combined complex tone as having a pitch of 100 Hz. Our ears seem to fill in the missing harmonic.

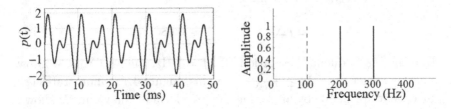

Figure 4.8: The waveform of the combination of 2 pure tones at 200 and 300 Hz with equal amplitudes, where the combined waveform has a period of 10 ms. Our ears perceive 100 Hz as the *virtual* or *residual pitch* due to a time period of 10 ms.

The periods of the 200-Hz and 300-Hz waves are 5 ms and 3.3333 ms, respectively. Because 10 ms is integer-divisible by both 5 ms and 3.3333 ms, the combined waveform appears to have a period of 10 ms (see Fig. 4.8). Since the temporal structure of the combined wave has a period of 10 ms, our ears seem to perceive the combined wave as having a period of 10 ms, and the pitch of the combined wave as being 100 Hz, which is the "absent fundamental" of 200 and 300 Hz. The combined wave has a *virtual pitch* of 100 Hz, in accordance with the "*temporal theory*" of pitch. The existence of virtual pitch causes a difficulty in the "*place theory*" of pitch because 200 Hz and 300 Hz tones do not excite the 100 Hz "location" of the cochlear membrane. The existence of virtual pitch favors the temporal theory of hearing because the auditory neuron pulses follow closely the time structure of an acoustic waveform and thus a perceived virtual pitch.

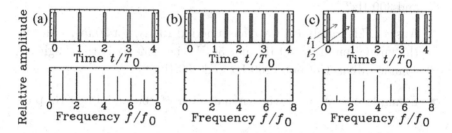

Figure 4.9: Seebeck's siren waveforms (top) and spectra (bottom): (a) Wave generated by wind passing through periodic equally spaced holes and the corresponding Fourier spectrum. Note that the pitch of $f_0 = 1/T_0$, where T_0 is the period of the air-pulses passing through the siren. (b) Wave for doubling the number of holes and the corresponding Fourier spectrum. Note that the pitch is $2f_0$. (c) Wave of displaced air pulses so that $t_1 + t_2 = T_0$, and the corresponding spectrum. The fundamental becomes small, *but* our pitch perception is f_0.

C.   Seebeck's Siren and Ohm's Law of Pitch

Following the Fourier theorem, Ohm proposed a theory that we perceive the tone of a complex sound by its frequency components, called Ohm's law of pitch. However, August Ludwig Friedrich Wilhelm Seebeck (1805–1849) carried out a series of periodic siren experiments to demonstrate that pitch perception is more complicated than frequency content. Figure 4.9 shows the periodic siren waveforms (left plots) and their corresponding Fourier spectra (right plots). The spectrum of the periodic pulse (a) has a period T_0. Its spectrum has harmonics nf_0, where $n = 1, 2, 3, \ldots$ and $f_0 = 1/T_0$. Our ears perceive the pitch as f_0. When Seebeck introduced additional siren holes evenly so that the period was reduced by a factor of 2, shown in pulse (b), the spectrum became $n \times 2f_0$, $n = 1, 2, 3, \ldots$. The pitch is twice the fundamental, i.e.

$2f_0$. When the siren holes were unevenly distributed with $t_1 + t_2 = T_0$, the dominant harmonic might be $2f_0$, while the amplitude of the fundamental frequency in the spectrum could be small. However, the perceived pitch would be f_0, i.e. we cannot identify pitch only by its Fourier spectrum. Seebeck attributed the pitch to the temporal waveform of the pulse, in contrast to Ohm's explanation.

★ Example 4.7: Why 50 Hz is not the virtual pitch of waves with 200 and 300 Hz?

Answer: The combined wave has a period 10 ms, not 20 ms, and thus the virtual pitch is 100 Hz, not 50 Hz, i.e. 100 Hz is the "highest" common divisor of both 200 Hz and 300 Hz.

★ Example 4.8: What is the virtual pitch of 400 and 1,000 Hz pure tones?

Answer: The highest common divisor of 400 and 1,000 is 200, thus the virtual pitch is 200 Hz.

★ Example 4.9: What is the perceived pitch of combining 400, 800, and 1,200 Hz pure tones?

Answer: The highest common divisor of 400, 800, 1,200 is 400, thus the virtual pitch is 400 Hz.

★ Example 4.10: Find the virtual pitch of pure tones of 500, 700, and 900 Hz played together.

Answer: The highest common divisor is 100, thus the virtual pitch is 100 Hz. For example, the wave form $p(t) = \cos(2\pi \times 500t) + \cos(2\pi \times 700t) + \cos(2\pi \times 900t)$, shown in the graph below, has a period of 10 ms. The virtual pitch is 100 Hz.

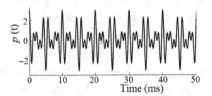
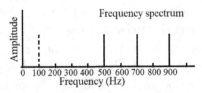

III. Theories of Pitch: Place vs Temporal Theory of Hearing

Research on hearing has made much progress in physiology since the pioneer work of G.v. Békésy. Single auditory neuro fiber recording shows clear frequency tuning, indicating the place theory for pitch. The synchronization between the recorded neuro-pulses and the external sound signals provides evidence of temporal analysis of our neuro-transmission network.

The place theory of hearing associates frequency with a resonant area of basilar membrane. Békésy provided support to the place theory in his experiments with cochlea removed from cadavers. Experiments with animals in vivo provide evidence of very sharp tuning of individual neuron fibers. Using Greenwood's formula, Fig. 3.3 shows the characteristic frequency vs position on the basilar membrane. Frequency maps of auditory neurons in the auditory system suggest the *place theory of the pitch*. Neuro-biological experiments showed further that the neurons in cochlear nuclei, the inferior colliculus, and auditory cortex have also divided into different pitch regions. The place theory can explain the effect of low-frequency sound masking high-frequency sound because the low-frequency sound sets up vibrations deeper in the cochlear membrane, while the high-frequency sound excites cochlear membrane closer to the oval window. However, recent experiments point to its limitation, e.g. the human ear having very fine frequency discrimination and our perception of the missing fundamental in a complex tone as the pitch, etc.

Evidence of the temporal (periodicity) theory of pitch is that the ear performs time analysis of a sound wave. Based on the Fourier theorem, the fundamental frequency of a periodic wave with period T_0 is $f_0 = 1/T_0$. The spectrum of the periodic wave is $f_0, 2f_0, 3f_0, 4f_0 \ldots$, shown in Fig. 4.10(a).

Depending on the envelope of the waveform, the amplitude of the fundamental at f_0 may be small or even zero as shown in Fig. 4.10(b). Our brain apparently perceives the (virtual) pitch of f_0. Our hearing system performs both time and frequency analysis of the sound wave and reaches a conclusion that the temporal information is more relevant.

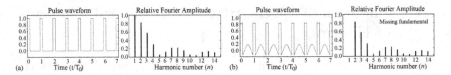

Figure 4.10: (a) Pulse waveforms and their frequency spectra with pulse width 19% of the period. (b) Waveform of (a) by removing the fundamental harmonic. The resulting waveform still maintains its periodicity, and the missing fundamental is the virtual pitch.

J.F. Schouten and his colleagues in the Netherlands performed experiments to support the temporal theory of pitch.[5] Their psycho-acoustic experiments used amplitude modulation of a pure tone with $p(t) = (1 + m \cos 2\pi g t) \sin 2\pi f t$, where $m = 0.9$, $g = 200$ Hz, and the carrier frequency f varied from 1,000 Hz to 2,000 Hz. Figure 4.11(a) shows $f = 1200$ Hz (top) and 1250 Hz (bottom). The top plot has a modulation frequency of 200 Hz because 1,200 Hz is divisible by 200 Hz. The bottom plot has a very weak amplitude modulation frequency of 50 Hz because the common divisor of 1,250 Hz and 200 Hz is 50 Hz. However, all subjects in the experiments perceived the virtual pitch to be 200 ± 40 Hz for all carrier frequencies f in 1,000–2,000 Hz, evidently agreeing with the temporal waveform of Fig. 4.11(a).

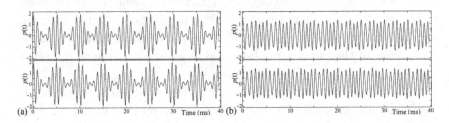

Figure 4.11: (a) The waveform of $p(t) = (1 + m \cos 2\pi g t) \sin 2\pi f t$, where $m = 0.9$, $g = 200$ Hz, and $f = 1,200$ Hz (top) and 1,250 Hz (bottom), used by Schouten *et al.* to demonstrate the virtual (residual) pitch. (b) The waveform of $p(t) = (m/2) \cos 2\pi (f - g)t + \sin 2\pi f t + (m/2) \cos 2\pi (f + g)t$ vs time, where the parameters are $m = 0.9$, $g = 200$ Hz and $f = 1,200$ Hz (top) and 1,250 Hz (bottom).

---

[5]See J.F. Schouten, R.J. Ritsma, and B.L. Cardozo, *JASA*, **34**, 1418 (1962).

It is worth pointing out that the waveform and pitch may depend
on the phases of these three harmonics. Figure 4.11(b) shows $p(t) =$
$\frac{m}{2} \cos 2\pi (f - g)t + \sin 2\pi ft + \frac{m}{2} \cos 2\pi (f + g)t$ vs time, where the parameters are $m = 0.9$, $g = 200$ Hz and $f = 1,200$ and $1,250$ Hz. Although
the Fourier amplitudes of Figs. 4.11(a) and 4.11(b) are identical, their
temporal waveform can be very different, and the virtual pitch of 200 Hz
may become weak in Fig. 4.11(b).

The temporal theory of pitch is further justified by recent advances
in neurophysiology. The neuron firing/impulse rate of the auditory
nerve depends on both the sound intensity I and the frequency f of
the sound — e.g. neurons do *not* fire on *every* oscillation cycle of frequency f for very faint sounds. Neurons *do* tend to fire on the peaks of
the waveform and resulting neuron signal in phase with the wave cycle.
They rarely fire at any other time. Figure 4.12 shows schematic neuron
signals of an individual neuron and the combination of all neuron signals
by a single tone. The spikes usually correspond to one or two periods
of the tone.[6] The combined signal of the summed auditory nerves syn-
chronizes with the peaks of a sinusoidal signal. Our brain can carry out
frequency analysis of the auditory nervous system firing pattern. The
central nervous system (CNS) can carry out autocorrelation analysis to
obtain the pitch. We perceive the loudness through the rate of auditory nerve fiber firing, the number of active neurons, and the firing of
high-threshold neurons.

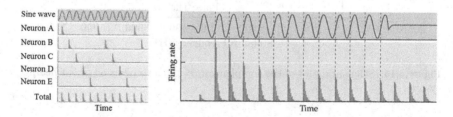

Figure 4.12: A schematic drawing of the excitation of a sine wave on the hearing
system. Each neuron may not fire at each peak of the wave, but combination of
many neurons provides the temporal structure of the sine wave.

---

[6]See e.g. I. Tasaki, *J. Neurophysiology*, **17**, 97 (1954); see also X. Wang, T. Lu,
R.K. Snider and L. Liang, *Nature*, **435**, 341 (2005).

Neurobiological experiments show that signals of some auditory neurons synchronize with the external signal up to about 3–5 kHz in the auditory nerve fiber. The time structure is nearly preserved through the neural transmission chain. The synchronization reduces along the auditory pathway and persists up to about 100 Hz at the auditory cortex in some mammals.[7] Warren and Wrightson carried out psychoacoustic experiments with iterated periodic noise signals by concatenating a noise signal of period $T_0/2$ with its inverse so that the resulting signal had a period of T_0.[8] People will perceive the pitch frequency as $2/T_0$ up to 100 Hz. Beyond 100 Hz, our brain recognizes pitch only as $1/T_0$.

Results of neuroscience research also show that "auditory neurons in humans are more sensitive to sound frequencies than most mammals." Furthermore, non-synchronized neurons can provide key information about pitch at higher frequencies. Tonotopic organization of brain areas mapped by using fMRI seems to support the place theory.[9] The place theory runs "out of room" for low-frequency information, and the temporal theory may have difficulty at high frequencies of neuron firing rate. It is likely that both place theory and temporal theory are operational in our hearing system. The temporal (or frequency) theory works for low frequency, and the place theory works at higher frequency. Both theories work together in the intermediate frequency region, as shown schematically in Fig. 4.13 below.

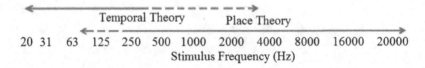

Figure 4.13: Schematic plot of the complementary pitch recognition of our hearing system.

---

[7]See Wang, X. Neural coding strategies in auditory cortex, *Hearing Research*, **229**, 81 (2007).

[8]See Fig. 3 in Richard M. Warren and John M. Wrightson, *JASA*, **70**, 1020 (1981). See also Chapter 3 in Ref. [20] for detailed surveys and discussions on pitch perception.

[9]See C. Humphries, E. Liebenthal, and J.R. Binder, *Neuroimage*, **50**, 1202 (2010); doi:10.1016/j.neuroimage.2010.01.046

IV.   Timbre and Tone Quality

Timbre is the attribute of auditory sensation by which a listener can judge two sounds with equal loudness and same pitch as different. Timbre depends primarily on the frequency spectrum, SPL, waveform, and the temporal characteristics (amplitude envelope). Timbre is subjective. Different people may perceive different aspects of the same sound signal.

A.   Fourier Analysis of Complex Tones and Fourier Spectrum in Music and Voice

Fourier analysis (or spectral or harmonic analysis) decomposes a complex waveform into its sinusoidal (sine and cosine) components. Each sinusoidal component has its amplitude, frequency, and phase. Table 4.1 shows components of higher harmonics of sawtooth, square and triangular periodic waves. Figure 4.14 shows waveforms of square, sawtooth, triangular, and pulse wave with pulse width of 19% of the period, and their associated spectra are at right.

Sawtooth wave: $\sin(x) - \dfrac{1}{2}\sin(2x) + \dfrac{1}{3}\sin(3x) - \dfrac{1}{4}\sin(4x) + \cdots$

Square wave: $\sin(x) + \dfrac{1}{3}\sin(3x) + \dfrac{1}{5}\sin(5x) + \dfrac{1}{7}\sin(7x) + \cdots$

Triangular wave: $\sin(x) - \dfrac{1}{3^2}\sin(3x) + \dfrac{1}{5^2}\sin(5x) - \dfrac{1}{7^2}\sin(7x) + \cdots$

Table 4.1: Amplitude and harmonic content of sawtooth, square, and triangular waves.

Wave	Harmonics	$|A_n/A_1|$	dB/octave
Sawtooth	All $(1, 2, 3, \ldots)$	$1/n$	$-6\,$dB
Square	Odd $(1, 3, 5, \ldots)$	$1/n$	$-6\,$dB
Triangle	Odd $(1, 3, 5, \ldots)$	$1/n^2$	$-12\,$dB

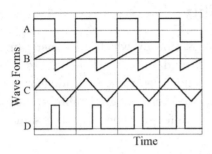 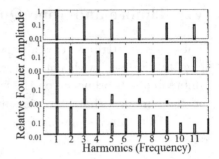

Figure 4.14: Some periodic waves (square, sawtooth, triangular, and pulse) waves vs their corresponding Fourier spectra. The time width of the pulse signal is 19% of the period. If the width of the pulse were 20%, the amplitude of harmonics 5, 10, 15, ... would be zero.

★ Example 4.11: The amplitude of the fundamental mode of a 250 Hz sawtooth wave is 3.0 volt. What is the amplitude of the next harmonic?

Answer: Frequency and amplitude of the next harmonics is 500 Hz and 1.5 V (see Table 4.1).

★ Example 4.12: If the fundamental of a square wave has an amplitude of 3.0 V at 250 Hz, what is the amplitude and frequency of the next harmonic?

Answer: The frequency and amplitude of the next harmonic are 750 Hz and 1.0 V (see Table 4.1).

A powerful Fast Fourier Transform (FFT) algorithm, judged as one of the TOP 10 algorithms in the 20th century (https://archive.siam.org/pdf/news/637.pdf), is available in most software packages. Real-time spectrum analyzers can produce real-time spectrum analysis. Figure 4.15 shows the longtime averaged power spectrum of a clarinet, piano, folk singer, radio newscaster, soprano, and tenor. Such information may be useful in determining the style of music, condition of instruments, and formants information of a singer. Longtime averaged spectra also depend on the written music, performance, room, and musical instruments. A reverse process of combining spectral components to form a complex tone or waveform is the Fourier synthesis.

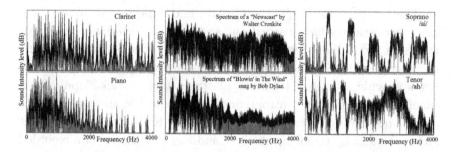

Figure 4.15: Spectra of musical instruments, a human speaking voice, and singing notes, providing pitch, note frequencies, and formants. The Fast Fourier Transform can analyze time sequences and provide diagnostics for music playing and singing styles.

B. Timbre and Dynamic Effects: Envelope and Duration and Vibrato

Timbre also depends on the transient frequency content and dynamics of the sound source. It is the key factor in recognizing musical sounds and personal speech. The envelope of a music note includes an attack, a steady state, and a decay portion. The playing backward of a recorded piano music may NOT sound like a piano at all. The timbre depends on the envelope of the sound. During a music attack, harmonics may develop differently in the transient. Figure 4.16 shows the wave and evolution of harmonics of a saxophone playing at 420 Hz. The harmonic content of an instrument may depend on the maker of the instrument and the player's control. The relative harmonic strength may differ at different times for the same player and same instrument.

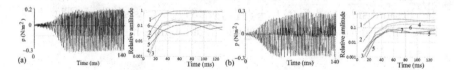

Figure 4.16: (a) The wave and the evolution of the first 7 harmonics of a saxophone playing a 420 Hz note during the attack 140 ms. (b) Similar plot for the same saxophone and same player playing the same note. Note that the spectrum of (a) has the highest second harmonic, while the spectrum (b) has the highest first harmonic.

The timbre scale is quite subjective and experiential. Since 1960, many acousticians have carried out systematic psychoacoustic experiments on the timbre of various musical instruments. Some instruments are easier to identify than others. Advancements in electronic synthesizers provide a toolbox for manipulating attack, attack slope, attack time, release, temporal centroid, duration, frequency, and amplitude of energy modulation, decay, and decay slope.[10] K.W. Berger carried out an experiment by removing the first and last half-second of a recorded sound of wind instruments (flute, oboe, clarinet, tenor sax, alto sax, trumpet, cornet, french horn, and trombone). Most of the 33 musician jurists could not identify their correct instruments. Strong and Clark preformed similar experiments and found that the oboe, clarinet, bassoon, tuba and trumpet are more sensitive to the spectrum; flute is more sensitive to the envelope; and trombone and French horn are sensitive to both spectrum and envelope. Most recent studies point to various dimensions of multi-dimensional scaling in timbre space.

C.   Modulation of Tone by Another

Vibrato is a rapid, slight variation of pitch that enhances music performance and contributes to timbre quality. It provides both frequency shifting (FM) and amplitude modulation (AM) to a musical performance at an average rate of 6–7 Hz in frequency modulation for most players. Amplitude modulation of a pure tone of frequency f_2 by a pure tone at frequency f_1 will produce a tone having spectrum f_2 and sidebands $f_2 \pm f_1$. For example, if we modulate a carrier frequency of 800 Hz with a 150-Hz signal, the spectrum of the modulated wave will include 650, 800, and 950 Hz. A nonlinear amplitude mixer will produce "additional sidebands" at $f_2 \pm 2f_1$, $f_2 \pm 3f_1$ ..., as shown in Fig. 4.17. Phase modulation will also produce all sidebands in the spectrum.

---

[10]G. Peeters, B.L. Giodano, P. Susini, N. Misdariis, and S. McAdams, *JASA*, **130**, 2902 (2011); K.W. Berger, *JASA*, **36**, 1888 (1964); W. Strong, and M. Clark, *JASA*, **41**, 277 (1967); T.M. Elliott, L.S. Hamilton and F.E. Theunissen, *JASA*, **133**, 389 (2013).

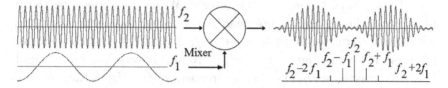

Figure 4.17: An example of amplitude modulation to a wave. The wave modulation has applications in wave transmission and wave decomposition.

D. Effect of Phase on Waveform and on Timbre

In 1877, Helmholtz found that "the quality of music tone depends solely on the number and relative strength of its harmonics." There is not much phase dependence. However, in 1969, Promp and Steeneken found that there were subjects who could easily differentiate the waveforms A and B from C and D in Fig. 4.18; while it was found to be much more difficult to differentiate between A and B; or C and D. Most people cannot distinguish the phase effects of a complex tone, BUT some people may have sensitivity to the phase of complex tone in timbre.

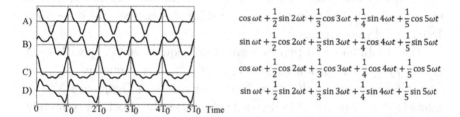

$$\cos\omega t + \frac{1}{2}\sin 2\omega t + \frac{1}{3}\cos 3\omega t + \frac{1}{4}\sin 4\omega t + \frac{1}{5}\cos 5\omega t$$

$$\sin\omega t + \frac{1}{2}\cos 2\omega t + \frac{1}{3}\sin 3\omega t + \frac{1}{4}\cos 4\omega t + \frac{1}{5}\sin 5\omega t$$

$$\cos\omega t + \frac{1}{2}\cos 2\omega t + \frac{1}{3}\cos 3\omega t + \frac{1}{4}\cos 4\omega t + \frac{1}{5}\cos 5\omega t$$

$$\sin\omega t + \frac{1}{2}\sin 2\omega t + \frac{1}{3}\sin 3\omega t + \frac{1}{4}\sin 4\omega t + \frac{1}{5}\sin 5\omega t$$

Figure 4.18: Four waveforms vs time. They have identical period and an identical Fourier spectrum. The angular frequency is $\omega = 2\pi f$. Some people may be able to differentiate these waves in their ears. Most people cannot hear the difference.

E. Nonlinear Effects on Timbre

Very loud sound can generate nonlinear response in all physical system, including our ears. The response R of our ears to the sound pressure p is $R = a_0 + a_1 p + a_2 p^2 + a_3 p^3 + a_4 p^4 + \cdots$, where a_n are constants that depend on the physical system. In the presence of a nonlinear response, we may hear all harmonics mf of the fundamental frequency, called "aural harmonics." Here m is an integer. When a complex tone with two frequencies f_1 and f_2 are heard in our ears, the aural harmonics will produce coupled harmonics at $|mf_1 \pm nf_2|$, where m and n are integers. The combination tone may produce a very low pitch sound in a small organ pipe. The tone at frequency $|f_1 - f_2|$ is the different tone or Tartini tone, after the Italian violinist Tartini, who reported the discovery in 1714. Musicians sometimes employ these difference tones in music performance.

V. Linear Superposition, Beat, Combination Tones, Consonance and Dissonance

The combination of two sine waves having the same frequency will generate a single sine wave at the same frequency, i.e.

$$A \sin\left(2\pi f t + \phi_A\right) + B \sin\left(2\pi f t + \phi_B\right) = C \sin(2\pi f t + \phi),$$

where the combined amplitude is $C = \sqrt{A^2 + B^2 + 2AB\cos(\phi_A - \phi_B)}$ depends on the relative phase, e.g. $C = A + B$, if $\phi_A - \phi_B = 0$; $C = \sqrt{A^2 + B^2}$ if $\phi_A - \phi_B = 90°$; $C = A - B$ if $\phi_A - \phi_B = 180°$, etc. The phase of two waves at the same frequency can interfere constructively at $\phi_A - \phi_B = 0$, called in-phase; or interfere destructively at $\phi_A - \phi_B = 180°$, called out-of-phase.

The combination of two sine waves at frequencies f_1, f_2 with small $\Delta f = |f_2 - f_1|$ will produce a sine wave at a "fusion frequency" $f_{\text{fusion}} = \frac{1}{2}(f_2 + f_1)$ with an *amplitude modulated by the "beat" frequency* Δf (see also Fig. 3.7). Figure 4.19 schematically shows the effects of combining two pure tones at small frequency difference.

If Δf is larger than the "fusion limit" F, the frequencies f_2 and f_1 can be distinguished. The beat disappears. The fusion limit F is about a semitone (or 100 cents) at 500 Hz; and the fusion frequency F is about a whole tone (200 cents) for $f < 200$ Hz or $f > 4$ kHz. For two tones with $\Delta f = |f_2 - f_1| < $ F, the resulting tone is fused into $f_{\text{fusion}} = \frac{1}{2}(f_2 + f_1)$. At $\Delta f < 10$ Hz, our ears hear the beat frequency Δf. The beat produced by combining two pure tones before reaching our ears is a physical effect, i.e. these two waves produce larger and smaller amplitude modulation vs time at the beat frequency. The beat may become "rough" to our ear at $\Delta f > 15$ Hz. The roughness in the

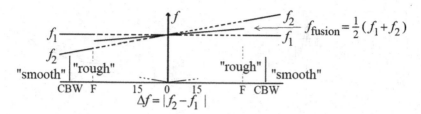

Figure 4.19: The superposition of two pure tones with small frequency difference can produce a fusion frequency at $f_{\text{fusion}} = \frac{1}{2}(f_2 + f_1)$ with amplitude modulation, called beat. The beat frequency is $f_{\text{beat}} = |f_2 - f_1|$. If the beat frequency is less than 10 Hz, we can hear beat. When the beat frequency is larger than 15 Hz, the combination tone is rough to our ears until the difference of frequency is higher than the "fusion frequency" F. If $|f_2 - f_1| > F$, then our ears perceive two separate tones.

tones may indicate that both frequencies activate the same part of the basilar membrane. The roughness disappears at a frequency separation equal to or larger than the critical bandwidth f_{CBW} of Eq. (3.8), where the two frequencies activate different sections of the basilar membrane. This phenomenon only applies to monaural listening with pure tones.

When two pure tones enter our left and right ears independently, we also hear "binaural beat," a phenomenon due entirely to the combination of the auditory neural transmission network. This means that our auditory nerve pulses retain the phases of sound waves, so that there is a modulation of higher or lower combined rates. Thus, we perceive variation of sound intensity at the beat frequency. Difficulty in perceiving the difference in pitch between the two ears is called diplacusis.

When we play two pure tones together, it may sound consonant (feeling pleasant) or dissonant (feeling unpleasant). Consonance is the combination of notes that sound pleasant to most people. Dissonance is the roughness to our ears when combining complex tones at some frequency difference. The question is, why are some combined tones more consonant than others? Is it genetic or biological evolutionary?

The harmonics of the standing wave spectrum of a string or wind instrument are integer multiples of the fundamental. All living things evolve with natural standing wave spectra in the environment. Human ears perceive consonant sound in tones having frequency ratios of simple integer fractions, e.g. octave (2:1); perfect fifth (3:2 G:C); perfect fourth (4:3 F:C); major sixth (5:3 A:C); major third (5:4 E:C); minor sixth

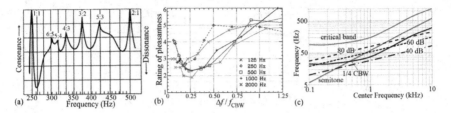

Figure 4.20: (a) Degree of consonance peaks at the small fractional-integer frequency ratio of two tones. (b) Degree of consonance vs the frequency difference. The most dissonance of the combined tone is the frequency difference at $\frac{1}{4}$ of the critical bandwidth. (c) The dissonance formula of Lameoka and Kuriyagowa at 80 dB (dashed), 60 dB (dotted), and 40 dB (dash-dot) is compared with the semitone and the $\frac{1}{4}$ critical bandwidth (solid).

(8:5 A^b:C); and minor third (6:5 E^b:C), all shown in Fig. 4.20(a). The combination of two pure tones at simple integer fraction ratios in frequencies is pleasant to our ears. Pythagoras designed such music tuning some 2,000 years ago.

Dissonance is a combination of notes that sound harsh or unpleasant to most people. Helmholtz pointed out that "when two tones create [a] beat frequency at 30–40 Hz, they are *rough* to our ears and they are dissonant."[11] The dissonance of two frequencies depends on the difference between the two tones and on the critical bandwidth. Figure 4.20(b) shows data on the "pleasantness rating" vs fraction of Δf to the critical bandwidth.[12] The most dissonant frequency difference occurs at about $\frac{1}{4}$ of the critical bandwidth of Eq. (3.8).

Lameoka and Kuriyagawa found a formula for greatest dissonance at the frequency difference $\Delta f_d = 2.27 \left(1 + \frac{L_p - 57}{40} \right) f^{0.447}$, where L_p is the sound pressure level and f is the frequency of the primary tone.[13] Figure 4.20(c) compares the dissonance frequency for $L_p = 40, 60, 80$ dB with $\frac{1}{4}$ of the critical bandwidth and the semitone interval.

[11]Helmholtz, H. von. (1877). *Die Lehre von den Tonempfindungen als physiologische Grundlage für die Theorie derMusik.* 1877, 6th ed., Braunschweig: Vieweg, 1913; trans. by A.J. Ellis as On the sensations of tone as a physiological basis for the theory of music (1885). Reprinted Dover, New York, 1954.

[12]See R. Plomp and W.J.M. Leveit, *JASA*, **38**, 548 (1965).

[13]Akio Kameoka and Mamoru Kuriyagawa, *JASA*, **45**, 1460 (1969).

Dichotic (or binaural) dissonance is also of great interest for acousticians and musicians. When two frequencies with a dissonance frequency difference activate our right and left ears separately, all people perceive less roughness. The combination of two consonance tones at a dissonance frequency shift is dissonance to our ears. However, in dichotic listening, i.e. they enter our left and right ears separately, and most people *do not* perceive dissonance, while some people may still perceive dissonance. The perception of dissonance is quite personal. Tones from musical instruments have many harmonics for a given note. Dissonances can occur between harmonics of consonant fundamental tones. Music making is also evolving. Musicians before the 19th century may have composed music more consonant to our ears, while modern musicians may compose music purposely dissonant to generate special sound effects to provoke our emotions.

VI. Pitch, Musical Scale and Temperament

Resonances exist in all physical systems. The spectra of many physical systems, such as strings, air columns, rods, etc., form integer multiples of a fundamental frequency. The frequency ratio of the 2nd harmonic to the fundamental is 2:1. Is it evolutionary or is it accidental that almost all musical scales choose the octave as the basic unit with frequency ratio 2:1? Regardless, our ears seem to recognize octave notes as the most consonant and similar, shown in Fig. 4.20(a).

Music scale is a link of notes arranged in ascending or descending order in the octave unit. The tuning is the adjustment of pitch in musical instruments so that they can be played together. The preference for tuning depends on the composer, musician and listener. Different cultures use different scales, e.g. the pentatonic or 5-tone scale in Chinese, Celtic and Native American music. A diatonic scale is an eight-note musical scale, composed of seven pitches (5 whole-steps and 2 half-steps) and a repeated octave, e.g. $(C_4*D_4*E_4F_4*G_4*A_4*B_4)C_5$ for the 4th octave. The two half-steps EF and BC are maximally separated from each other. The asterisks between two notes in the scale sequence represent either the *sharp* or *flat* of adjacent notes for 12 semitones in 1 octave. Each step is a semitone. According to studies, the twelve-tone musical scale may have its roots in the sound of the human voice during the course of evolution.[14] Analysis of recorded speech samples found peaks in acoustic energy that mirrored the distances between notes in the twelve-tone scale. Human voices have high probability in amplitude-frequency combinations that are chromatic and consonant. Today, the relationship between vocal and music scale is still a very active subject of research in human language and music. Figure 4.21 shows the standard piano 88-key keyboard and the music notation of the 4th octave in diatonic scale.

Figure 4.21: Standard piano 88-key keyboard from A_0 octave to C_8 and the music staff symbol for the 4th octave.

The pitch standard set the 4th octave A-note, denoted by A4, at 440 Hz in 1939 during an international conference in London. The 10 octaves in our hearing range are C_0 (16.35 Hz) to C_{10} (16,744 Hz). The assignment of frequency ratio varies on different temperament arrangements. The Indian music scale appears very complicated, but it was pointed out by A.H. Benade (see Ref. [23]) that Indian music uses a 7-tone scale similar to the major scale of Western music. The microtonal decoration is only a manner of performance style. Table 4.2 lists various tuning methods of the 12-notes of the 4th octave. The Raga North Indian scale divides an octave into 22 steps.[15] We list its 12 major steps in Table 4.2.

---

[14]See e.g. Schartz, Howe and Purve, The statistical structure of human speech sound predicts musical universals, *J. Neuroscience*, **23**, 7160 (2003).

[15]See e.g. http://www.ancient-future.com/hindustaniraga.html

Table 4.2: The scale of major temperaments along with the North Indian raga scale.

Name	Equal temperament	Ratio	Just diatonic	Pythagorean	Music name	North Indian
C_4	261.63	1	$1 = 4/4$	1	Unison	1
$C_4\#/D_4b$	277.18	1.0595			Minor 2nd	1.067
D_4	293.66	1.1225	$9/8$	$9/8 = 1.125$	Major 2nd	1.125
$D_4\#/E_4b$	311.13	1.1892			Minor 3rd	1.2
E_4	329.63	1.2599	$5/4 = 1.25$	$81/64 = 1.2656$	Major 3rd	1.25
F_4	349.23	1.3348	$4/3$	$4/3$	Major 4th	1.333
$F_4\#/G_4b$	369.99	1.4142			Diminished 5th	1.39
G_4	392.00	1.4983	$6/4 = 3/2$	$3/2$	Perfect 5th	1.5
$G_4\#/A_4b$	415.30	1.5874			Minor 6th	1.6
A_4	440	1.6818	$5/3 = 1.6667$	$27/16 = 1.6875$	Major 6th	1.667
$A_4\#/B_4b$	466.16	1.7818			Minor 7th	1.775
B_4	493.88	1.8877	$15/8 = 1.875$	$243/128 = 1.8984$	Major 7th	1.875
C_5	523.25	2.0000	2	2	Octave	2

★ Example 4.13: Explain the meaning that music pitch obeys Weber–Fechner's law.

Answer: The basic unit in most musical scales is the octave at the frequency ratio of 2:1. Music pitch obeys Weber–Fechner's law, i.e. $f_{C5} = 2f_{C4} = 2^2 f_{C3} = 2^3 f_C = 2^4 f_{C1} = 2^5 f_{C0}$. The frequency ratio of the adjacent note is a constant. In different cultures, music scales divide the octave in different ways, while the frequency ratio of an octave is always 2. This is Weber–Fechner's law.

A. The Pythagorean Tuning

A temperament is a system of tuning in which the intervals is most pleasing to musicians and listeners. The consonance or dissonance of notes depends on their frequency ratio. The intervals which are perceived to be most consonant are composed of small integer ratios of frequencies, i.e. at the ratios of 2:1 (octave); 3:2 (fifth); 4:3 (fourth); 5:4 (major third); and 5:3 (major sixth), shown in Fig. 4.20. This small-integer ratio is the basis of Pythagorean tuning and just tuning. Pythagoras (570–495 BC) has been given credit for dividing the string into segments of halves, thirds, and fourths that produce harmonic tones. The Pythagorean musical scale emphasizes the major 4th (F) and 5th (G) at the frequency ratio of 4/3 and 3/2, shown in Fig. 4.22(a).

The major 3rd (E) note is derived from C with frequency ratio: $\frac{1}{4}(\frac{3}{2})^4 = \frac{81}{64}$ and the major 6th (A) is the frequency ratio $\frac{1}{2}(\frac{3}{2})^3 = \frac{27}{16}$. The major 2nd (D) is the major 4th of G-note or the major 5th of A-note. The major 7th (B) is the major 3rd of G-note. This completes all major notes listed in the fifth column in Table 4.2. Ambiguity begins when one try to fill the minor notes, e.g. the $G^{\#}(A^b)$-notes derived from C and E has different frequencies as shown in Fig. 4.22(a). This problem persists when one tries to fill 12-notes in the diatonic scale, clearly shown in Fig. 4.22(c).

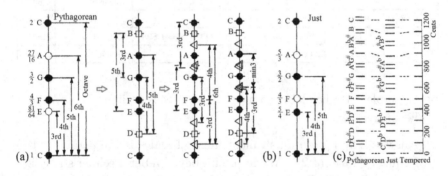

Figure 4.22: (a) The Pythagorean tuning scheme is based on the 4th and 5th at frequency ratio of 4/3 and 3/2. All 8-notes in an octave has simple frequency ratios. Complication occurs when we expand the scale to 12-note diatonic scale. (b) Just tuning emphasizes the frequency ratio of 4:5:6 for C:E:G. Similar difficulty will appear in just tuning. (c) The complication and ambiguity of the 12-tone diatonic scale for the Pythagorean and just tunings, and the unification of these 12 notes into the equal temperament tuning.

B.   The Just-Intonation Tuning

Just tuning emphasizes the major 3rd (E) and 5th (G), or is based on major triads C:E:G with frequency ratios of 4:5:6, as shown in Fig. 4.22(b). The 4th column in Table 4.2 shows the just intonation scale. Unfortunately, it also produces ambiguity in major and minor tone intervals in the diatonic 12-note scale. The great disadvantage of the Pythagorean scale is poor tuning of the 3rd, e.g. the E/C of 81/64 is larger than $5/4$ of the just tuning by a factor 1.0125, which is beyond the jnd (see Fig. 4.3). Although a "meantone temperament" tries to make corrections to large frequency deviations, the meantone becomes out of tune for sharps and flats. Figure 4.21(c) shows similar non-uniqueness when we try to fill the diatonic 12-note scale. The equal temperament intended to unify the ambiguity of the diatonic 12-note music scale.

C.   The Equal Temperament Tuning

The equal temperament scale is a compromise for music tuning, and is the most common tuning system in Western music since the late 18th century. Zhu Zhaiyu, a prince of the Ming-dynasty in China, has been credited for the invention of the equal temperament in his 500-page book *Fusion of Music and Calendar* published in 1580.

The equal temperament scale is based on five whole tones and two semitones. Each semitone is increased by $2^{1/12} = 1.05946$. A whole tone is twice a semitone, i.e. increasing the frequency by a factor of $2^{2/12} = 1.12246$. Each semitone in equal temperament scale is divided into 100-cents for a total of 1200 cents in one octave, or 1 cent has the frequency ratio of $2^{1/1200} = 1.0005777895$. Musical instrument makers adopt the equal temperament in their design of fretted instruments or the tone hole of wind instruments. However, musicians may use other tunings in the playing of non-fretted musical instruments.

★ Example 4.14: In equal temperament, what is the increment factor of a semitone? What is the increment factor for 1 cent?

Answer: Western music divides the octave into 12 intervals called semitones, with names A through G, including sharps and flats, as shown in the graph for the 4th octave. In equal temperament, each semitone corresponds to frequency ratio of $2^{(1/12)} = 1.05946$. For example, $f_{A\#} = f_A \times 1.05946$, $f_{Ab} = f_A/1.05946$, etc. Similarly, $f_{C_5} = f_{A_4} \times (1.05946)^3$ because C_5 is 3 semitones above A_4, and $f_{E_4} = f_{A_4}/(1.05946)^5$ because E_4 is 5 semitones below A_4. Each semitone interval has 100 cents, where a 1-cent increment is a frequency increment by the factor $2^{(1/1200)} = 1.000578$.

D.   Other Tuning Methods and Problems in Musical Scale Tunings

The diatonic scale is not the universal adopted music tuning method.[16] Other tuning methods include the N-note equal temperament scale, where the adjacent music notes have a frequency ratio of $2^{(1/N)}$. The diatonic equal temperament scale has N = 12. Other popular scales have N = 15, 19, 22, 31, and 53, etc. The fretboards of fretted string instruments are spaced according to the equal tempered tuning of various scale arrangements.

The tone holes in most wind instruments are normally spaced according to the diatonic equal temperament tuning. However, the resonance frequencies depend on the speed of the sound wave (or the air temperature) in wind instruments. The frequencies of music notes may vary during the performance as the temperature of the wind instrument increases. The jnd of our ears is about a 3°C temperature change (see Examples 4.1 and 4.18).

There are fretted or non-fretted (fretless) string instruments. For fretted string instruments, the fretboards are spaced according to the diatonic equal temperament tuning. For non-fretted instruments, such as violins, violas and cellos, the open strings are normally tuned in perfect fifths. For example, the violin has open strings tuned to G3–D4–A4–E5, and violas and cellos have open strings tuned to C3–G3–D4–A4 based on the perfect fifth at frequency ratio 3/2. This suggests that their semi-tone ratio is slightly higher than in the conventional twelve-tone equal temperament. Because a perfect fifth has a 3:2 ratio in its base tone, and this interval is covered in 7 steps, each semi-tone has a frequency ratio of $(3/2)^{1/7} = 1.059634 = 2^{(100.28/1200)}$, i.e. 100.28 cents. Similarly, the octave will have a frequency ratio of $(3/2)^{12/7} \approx 2.00388{:}1$, instead of a 2:1 ratio. However, violinists choose pitches by ear during actual play, and only the four open strings have this 3:2 ratio. For diatonic music tuning, musicians may choose to play or perform in their preferred intonation. Studies found that many performers deviate from equal temperament and lean toward Pythagorean

---

[16]https://en.wikipedia.org/wiki/List_of_musical_scales_and_modes

intonation. Many choral conductors prefer a slightly higher major-third, especially in sustained chords and cadences.

The piano (A_0–C_8 with frequency range from 27.5 to 4186 Hz) is normally tuned based on the equal temperament. Figure 4.23 shows the O.L. Railsback curve. It appears that piano tuners tend to stretch the low and high octaves by up to 30 cents. Pianos were typically tuned by the overtone beat of a note with a reference note of the adjacent strings. The successive overtones are higher than they 'should' be [Eq. (2.15)], due mainly to the inharmonicity in the strings.[17] The Railsback curve is mostly within ±10 cents, the jnd of human ears (see Example 4.17).

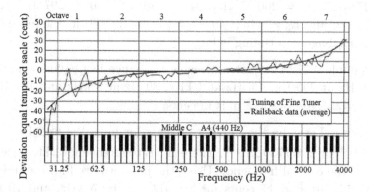

Figure 4.23: Railsback curve shows the systematic deviation of piano equal tempered tuning.

★ Example 4.15: The diatonic scale of Western music divides the octave into 12 steps called semitones e.g. C*D*EF*G*A*BC. The standard tone of the A-note at the fourth octave is set at 440 Hz. Find the frequencies of the following notes: B_2^b _____ Hz; $F_3^\#$ _____ Hz; D_5^b _____ Hz.

Answer: Note that A_2, A_3, and A_5 are 110 Hz, 220 Hz, and 880 Hz respectively. Since B_2^b is one semitone above A_2, we find $B_2^b = 110 \times 2^{1/12} = $ __116.5__ Hz; $F_3^\#$ is 3 semitones below A_3, $F_3^\# = 220/2^{3/12} = $ __185__ Hz; D_5^b is 8 semitones below A_5, $D_5^b = 880/2^{8/12} = $ __554__ Hz.

---

[17]See R.W. Young, Inharmonicity of plain piano string, *JASA*, **24**, 267 (1952).

★ Example 4.16: Violin open strings are tuned according to the Pythagorean tuning for $G_3 = 440 \times (2/3)^2 = 195.56$ Hz; $D_4 = 440(2/3) = 293.33$; $A_4 = 440$ Hz; and $E_5 = 440 \times (3/2) = 660$ Hz. What are the frequencies of these four open strings, tuned according to equal temperament and just intonation?

Answer: When tuned by equal temperament, their frequencies would be $G_3 = 220/2^{2/12} = 196.00$ Hz; $D_4 = 440/2^{7/12} = 293.66$ Hz; $A_4 = 440$ Hz; and $E_5 = 440 \times 2^{7/12} = 659.26$ Hz. Their difference from the Pythagorean tuning is within 1 Hz. They are all within our jnd. When tuned by just intonation, their frequencies are $G_3 = 297 \times 27/40 = 200.47$ Hz; $D_4 = 440 \times 27/40 = 297$ Hz; $A_4 = 440$ Hz; and $E_5 = 440 \times 40/27 = 651.85$ Hz. Their difference from the Pythagorean tuning is large.

★ Example 4.17: The just noticeable difference in frequency is about 2.5 Hz at 200 Hz, 4.5 Hz at 1 kHz, 30 Hz at 5 kHz and 100 Hz at 10 kHz. Express these jnds in cents.

Answer: Let x be the number of cents for jnds. We find $1 + \frac{\Delta f}{f_0} = 2^{x/1200}$ or $x \approx 1730 \Delta f/f_0$, where Δf is the frequency difference and f_0 is the central frequency. Using the jnd data, we find the jnds are 22, 7.8, 10, and 17 cents for 200 Hz, 1 kHz, 5 kHz, and 10 kHz, respectively. There is a saying: "You can hear about a nickel's worth of difference, i.e. jnd $\sim$ 5 cents." This calculation finds a *"dime's difference"* for about 10 cents. The Railsback curve is within the jnd of human ears.

★ Example 4.18: Many acoustic experiments use the octave band noise signal around center frequencies at 31.5, 63, 125, 250, 500, 1,000, 2,000, 4,000, 8,000, 16,000 Hz. Find the frequency range of $\frac{1}{3}$ octave band around 1,000 Hz.

Answer: The $\frac{1}{3}$ octave band around the 1,000 Hz is $(2^{-1/6} \times 1,000, 2^{1/6} \times 1,000) = (891, 1122)$ Hz. The ratio between the high and low cutoff frequencies is $2^{1/3}$, thus called $\frac{1}{3}$ octave band.

VII.   Summary

This chapter discussed two attributes of sound: pitch and timbre. Pitch is "the attribute of auditory sensation in terms of which sounds may be ordered on a scale extending from low to high." Pitch depends essentially on the frequency of the tone. In 1939, an international conference in London adopted 440 Hz as A4, now universally adopted by musicians and musical instrument makers. We recognize the pitch of a sound mainly through either the period of the time structure or the frequency content of the sound.

Timbre is the attribute of auditory sensation by which a listener can judge two sounds that are similarly presented and that have the same loudness and pitch as being dissimilar. Timbre depends primarily on the frequency spectrum, SPL, waveform (attack, duration, decay), and temporal characteristics (amplitude envelope).

A complex sound is a combination of sinusoidal sounds at various frequencies. A periodic wave is composed of harmonics that are integer multiples of the fundamental. The Fast Fourier Transform algorithm is available in almost all computer packages and can analyze the spectrum of a sound source in real time.

Physiological experiments imply that pitch recognition is associated with position on the cochlear membrane (place theory). Neurophysiological experiments also identify pitch as the temporal structure of the neuron pulses transmitted through the cochlear nucleus, superior olivary nucleus, inferior colliculus, and medial geniculate nucleus to our auditory cortex (temporal theory). The tuning curve of auditory neurons on the cochlear membrane is sharp in frequency, and the transmitted neural pulses maintain their synchronization in the transmission chain. Our auditory cortex is tonotopically organized. The interplay between the place and temporal theories is important in our perception of pitch. Because of the temporal structure of the neural pulses, we perceive a virtual pitch of the sound signal.

Linear superposition of two pure tones with nearly the same frequency produces beats at $|f_1 - f_2|$ and nonlinear beats at $|mf_1 - nf_2|$, with integers m and n. The nonlinear harmonics, called aural harmonics, are an indication that our auditory system is nonlinear. Musicians sometimes employ these difference tones in musical performances.

When combining two tones f_1 and f_2 together, a lower combination tone $|f_1 - f_2|$ is frequently heard. This is called a different tone or Tartini tone, after the Italian violinist Tartini, who reported the discovery in 1714. When the frequency difference is below 15 Hz, we hear a single tone at the fusion frequency with amplitude modulation at the beat frequency Δf. As the frequency difference increases, a sensation of roughness or dissonance develops. Psychoacoustic experiments show that the highest degree of dissonance occurs at $\Delta f \sim \frac{1}{4} f_{\mathrm{CBW}}$, where f_{CBW} is the critical bandwidth. The dissonance may depend on psychoacoustics, physiology, culture, and the "mood and personality" of the person at the time of listening. When the frequency difference is greater than a fusion frequency limit, we can simultaneously hear two tones. The degree of pleasant is highest when the ratio of frequencies of two tone is a small fractional-integer ratio. In particular, we judge that the tones with frequency ratio of 2:1 as similar and most consonance. The tone at frequency ratio of 2:1 is an octave. The octave is the basic unit in the musical scale in all cultures. This may arise from evolution of human hearing system because the resonant frequencies of natural systems have small fractional integer ratio (see Sec. IV in Chapter 2).

The use of the octave, with its 2:1 frequency ratio, as the basic unit in the music scale in all cultures may arise from the evolution of the human hearing system. Music tuning is the subdivision of the octave into scale tunings. Different cultures have developed their music scales through history. Although there are many music tuning methods, we discussed four major scale construction methods: Pythagorean tuning, meantone temperament, just tuning, and equal temperament. Each has its advantages and disadvantages. Most musicians popularly adopt the equal temperament, which divides the octave into twelve equal tempered semitone intervals of $2^{1/12} = 1.05946$. Each semitone is subdivided into 100 cents, with each cent having the frequency ratio of $2^{1/1200} = 1.000578$. Table 4.3, reproduced from Wikipedia, shows various tempered music scales, historically implemented in various cultures.

The frequencies of the 4th octave notes in equal temperament are shown in the Table 4.4. The 5th column shows the logarithm of the frequency ratio, wherein the intervals are all equal, indicating the psychoacoustic scale of our hearing system.

Table 4.3: Historical evolution of the equal tempered tuning reproduced from the Wikipedia.

YEAR	Name of the inventor	RATIO	CENTS
400	He Chengtian	1.060070671	101.0
1580	Vincenzo Galilei	18:17 [1.058823529]	99.0
1581	Zhu Zaiyu	1.059463094	100.0
1585	Simon Stevin	1.059546514	100.1
1630	Marin Mersenne	1.059322034	99.8
1630	Johann Faulhaber	1.059490385	100.0

Table 4.4: The frequency and frequency ratio of the 4th octave for the equal tempered musical scale. The logarithm of the frequency ratio in the 5th column has equal spacing, i.e. it obeys the Weber–Fechner's psychoacoustic law, meaning equal distance between successive notes.

Name	Equal temperament	Ratio	Name of notes	Log(Ratio)/Log(2)
C5	523.25	2.0000	Octave	1.000000
B4	493.88	1.8877	Major 7th	0.916667
A#/Bb	466.16	1.7818	Minor 7th	0.833333
A4	440	1.6818	Major 6th	0.750000
G#/Ab	415.30	1.5874	Minor 6th	0.666667
G4	392.00	1.4983	Perfect 5th	0.583333
F#/Gb	369.99	1.4142	Diminished 5th	0.500000
F4	349.23	1.3348	Major 4th	0.416667
E4	329.63	1.2599	Major 3rd	0.333333
D#/Eb	311.13	1.1892	Minor 3rd	0.250000
D4	293.66	1.1225	Major 2nd	0.166667
C#/Db	277.18	1.0595	Minor 2nd	0.083333
C4	261.63	1	Unison	0.000000

VIII.   Homework 04

1. Human hearing covers 10 octave bands in frequencies from 20–20,000 Hz. The center frequencies of the standard octave bands are 31.5, 63, 125, 250, 500, 1000, 2000, 4000, 8000, and 16,000 Hz. The lowest frequency and the highest frequency of these octave bands are _____ Hz and _____ Hz.

2. In western music, the *equal temperament tuning* subdivides the octave into 12 intervals called semitones. An octave begins with C and end with B with scale C*D*EF*G*A*BC to the next C-octave, where * symbol represents semitone between the whole tone. Since A4 is set at 440 Hz, the frequency of each semitone is obtained by the factor of $2^{1/12}$. For example counting F4 is 4-semitone *below* A4, its frequency is $440/2^{(4/12)} = 349$ Hz. The D5 note is 5-semitone *above* A4, its frequency is $440 \times 2^{(5/12)} = 587$ Hz. Thus the E4 frequency is _____ Hz. The frequency of C5 is _____ Hz, the frequency of $F_3^{\#}$ is _____ Hz, and $B_2^{\flat}$ is _____ Hz.

3. The just noticeable difference in frequency (pitch limen) is $\Delta f/f = 0.007$ above 500 Hz. This corresponds to _____ cents.

4. Using the Greenwood CF formula in Eq. (3.1) to find the position, where the characteristic frequency is 2000 Hz; $x =$ _____ .

5. _____ synthesis involves the creation of a complex tone or waveform by combining its spectral components in voice. Fast _____ transform is an algorithm developed in the 20th century that can extract of the frequency contents of a waveform.

6. Based on the uncertainty principle, the best frequency resolution that a Fourier analysis can achieve based on the sampling time window of 10 ms; 50 ms; and 100 ms, are _____ Hz, _____ Hz, and _____ Hz.

7. A complex waveform repeats itself every 4.0 ms. The frequencies of the first three Fourier components are _____ Hz, _____ Hz and _____ Hz.

8. The fundamental pitch of human voice is 100 Hz for men, 200 Hz for women, and 300 Hz for children. This pitch is created by air-pulses from the opening and closing of the vocal folds. What are the periods of human vocal air-pulses? _____ ms for men, _____ ms for women and _____ ms for children.

9. The virtual pitch of 3 pure tones 300 Hz, 500 Hz, and 700 Hz playing together? _____ Hz.

10. What is the virtual pitch of 3 pure tones 400 Hz, 600 Hz, and 800 Hz? _____ Hz.

11. What is the virtual pitch of 3 pure tones, 750 Hz, 1,000 Hz, and 1,250 Hz? _____ Hz.

12. Fourier analysis of a square wave indicates that its spectrum is composed of harmonics of (odd/even) order (choose one).

13. The frequencies of the first three harmonics of a 300 Hz square wave are _____ Hz, _____ Hz, and _____ Hz. If the amplitude of the fundamental is 1.00 A, then the amplitudes of the second and third harmonics are _____ A, and _____ A.

14. If the amplitude of the fundamental for a certain sawtooth wave is 2.0 volts, what are the amplitudes in volts of the next three harmonics? _____, _____, and _____. What are the harmonic numbers? _____, _____, and _____.

15. If the amplitude of the fundamental for a certain triangular wave is 3 volts, what are the amplitudes in [volts] of the next three existing harmonics? _____, _____, and

_____. What are the harmonic numbers? _____,
_____, and _____.

16. The intensity level of higher harmonics decreases by _____
dB/octave for the square wave; _____ dB/octave for the sawtooth wave, and _____ dB/octave for the triangular wave.

17. The pulse shape of human vocal air-pulses resembles the triangular wave. Thus the intensity level of harmonics decreases by
_____ dB/octave.

18. Amplitude modulation of 700 Hz pure tone with a 120 Hz pure
tone will produce a complex tone with frequencies _____ Hz,
_____ Hz, and _____ Hz.

19. Two square waves of frequencies 70 Hz and 200 Hz are played
together. The smallest beat frequency expected is _____ Hz.

20. Will two sine waves of frequencies at 70 Hz and 200 Hz sounded
together produce a beat frequency at 10 Hz? _____ (Y/N). Will
two triangular waves of frequencies at 70 Hz and 200 Hz sounded
together produce a beat frequency at 10 Hz? _____ (Y/N)

21. _____ is "rough" to our ear that results from two tones with
appropriate frequency difference are combined simultaneously.

22. A piano has 88 notes grouped in 12 semitones per octave. The
modern tune standard has the note A4 = 440 Hz. (a) If the piano
is tuned to the scale of equal temperament, what is the frequency of the middle C or $C_4 =$ _____ Hz; and the
note $G_4 =$ _____ Hz. (b) What is the ratio of frequency:
$G_4/C_4 =$ _____. How is the ratio compared with the perfect fifth ratio of 3:2?

23. Our ability to discriminate pitch, the just-noticeable difference
(jnd or Δf), is about 0.5% of the pure tone frequency. At a frequency of A4, or 440 Hz, this corresponds to $\Delta f =$ _____ Hz,

or _____ cents. Since the semitone spacing at A4 is _____ Hz, the fraction of the semitone spacing we can distinguish at 440 Hz = _____ .

24. What are the frequencies of the open strings of a violin, tuned according to the Pythagorean tuning? G3: _____ Hz; D4: _____ Hz; A4: 440 Hz; and E4: _____ Hz. What is their frequency difference from the equal temperament tuning?

25. What are the frequencies of the open strings of a viola, tuned according to the Pythagorean tuning, i.e. they are in all *perfect fifth* intervals? C3: _____ Hz; G3: _____ Hz; D4: _____ Hz; and A4: 440 Hz. What is their frequency difference from the equal temperament tuning?

26. What are the frequencies of the open strings of a cello, tuned according to the Pythagorean tuning, i.e. they are in all *perfect fifth* intervals? C2: _____ Hz; G2: _____ Hz; D3: _____ Hz; and A3: 220 Hz. What is their frequency difference from the equal temperament tuning?

27. What are the frequencies of the open strings of a double bass, tuned according to the Pythagorean tuning, i.e. they are in all *major fourth* intervals? E1: _____ Hz; A1 = 55 Hz; D2: _____ Hz; and G2: _____ Hz. What is their frequency difference from the equal temperament tuning?

28. Using the equal temperament tuning to find the frequency of the lowest open notes of brass instruments: trumpet B3-flat _____ Hz, trombone B2-flat _____ Hz, French horn F2 _____ Hz, and tuba B1-flat _____ Hz.

29. A pure tone with a frequency f_1 = 800 Hz is modulated by a frequency f_2 = 160 Hz. The primary sidebands are _____ Hz and _____ Hz. The virtual pitch heard will be the related fundamental, _____ Hz.

30. The Railsback curve (see Fig. 6.3) shows that piano tuners tend to stretch the low and high octaves by up to 30 cents away from the equal temperament tuning. Please explain the reason of this effect.

31. What would be the frequencies of the lowest and highest notes of 88-key piano based on the equal temperament tuning? What are the frequencies of A0 and C8 if the A0 is tuning downward by 40 cents, and C8 note is tuning upward by 30 cents? Are these frequency differences within the corresponding just noticeable differences?

32. Critical bands act like data collection units on the basilar membrane and are related to auditory sensations such as pitch and timbre. Using Eq. (3.8) to the critical bandwidth at central frequencies 100 Hz, 1,000 Hz, and 10,000 Hz: _____ Hz, _____ Hz and _____ Hz.

33. Human ears can hear sound covering frequencies from 20 Hz to 20,000 Hz. The standard center frequencies of the 10 octave bands are

f_c (Hz)	31.5	63	125	250	500	1000	2000	4000	8000	16000

Find the frequency ranges of the octave bands centered at 31.5 Hz, and 16,000 Hz.

34. The equivalent rectangular band (ERB) works as filter function of the cochlear membrane. Using the ERB in Eq. (3.9) to find the ERBs at center frequency 100 Hz, 1,000 Hz, and 10,000 Hz: _____ Hz, _____ Hz, _____ Hz.

35. Find the frequency of a pair of semitones that they are the most dissonant, based on the $\frac{1}{4}$ CBW condition. _____ Hz and _____ Hz.

36. Based on Fig. 4.20, the pair of semitones crosses the $\frac{1}{4}$ CB at about 300 Hz. The pair is most dissonant. At what frequency will the pair of whole tones be the most dissonant based on the $\frac{1}{4}$ CB criterion?

Chapter 5

Basic Physics of Some Musical Instruments

The science of musical instruments and their classification is "organology" or "instrumentology." Musical instruments evolved with time and are intimately related to culture. The basic structure of all musical instruments is a *sound source* and a *resonant body* that enhances sounds at particular frequencies. Sound sources (vibrational media) include the violin string, drum membrane, wood or metallic marimba bar, and the air stream in wind instruments. Resonance bodies include sound boxes for string instruments and drums, pipes for brass and wind instruments, etc. The design of musical instruments is part art, part engineering, and some basic physics. The aim is to provide excellent sound quality, aesthetic looking and easy to play and control. Modern electronic synthesizers can mix different sinusoidal tones to produce special electronic sounds that can emulate all kinds of musical instruments. The field of musical instruments is broad; we discuss only some basic physics principles of string and wind instruments and resonances of some percussion instruments.

I. String Instruments

The sound source of string instruments is the vibrating string. The standing waves or resonances on the string provide pitch of the note. Resonances of the sound box provide the characteristics of the music note. The wave speed on the string is $v = \sqrt{F_T/\mu}$, where $\mu = M/L$ is the mass per unit length and F_T is the tension [see Eq. (2.4)]. The resonance frequencies of standing waves of a string of length L, fixed at both ends, are $f_n = n\frac{v}{2L}$, where $n = 1$ is the fundamental mode,

141

and $n = 2, 3, 4 \ldots$ are harmonics. The string instruments are played by bowing, plucking, strumming, or hammering. They will produce different timbre by generating different harmonic content and pulse shapes. The sound boxes of string instruments are the dominating factor in their acoustic properties.

Historically, string instruments come in shapes and sizes as shown in Fig. 5.1. They will produce different timbre by generating different harmonic contents and pulse shapes. The sound boxes of string instruments are the dominating factor in their acoustic properties. Several renowned institutions devote their research to the design and characterization of string instruments.

Figure 5.1: Sting instruments comes in shapes and sizes from different cultures.

A.   Harmonics of Plucked and Bowed Strings

The harmonic content of a music note depends on how we play the instrument. For example, Figure 5.2(a) shows the time evolution of wave on a string plucked at its mid-point and its corresponding spectrum. The spectrum has missing even harmonics. Since the mid-point of the string *must* be a "node" for all even harmonics (see Fig. 3.10),

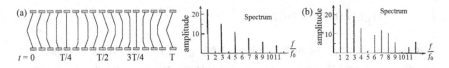

Figure 5.2: (a) Schematic drawing of the wave motion in 1 period of a plucked string at the middle. The spectrum does not have even harmonics because waves of all even harmonics must have a node at the mid-point. Since the string is plucked at the mid-point, even harmonic are not allowed. (b) The spectrum of a string plucked at $\frac{1}{5}$ of the string length. The spectrum will have missing 5, 10, 15 ... harmonics.

and the string plucked at mid-point *cannot* be a "node," all even harmonics are missing. Similarly, Fig. 5.2(b) shows the spectrum of a string plucked at $\frac{1}{5}$ of the string length. The spectrum will have missing 5, 10, 15 ... harmonics because these harmonics require a node point at the $\frac{1}{5}$ position of the string.

When bowing a string, the string stick to the bow until the friction force is smaller than the restoring force. This sets up a sawtooth wave motion on the string, as shown schematically in Fig. 5.3(a). One generally has all harmonics in the violin spectrum. Figure 5.3(b) shows the spectra of bowed violin open string G and D notes. The enhancement of certain harmonics is related to the resonances of the sound box of the instrument. In general, the acoustic properties of a string instrument can be determined by carrying out systematic measurements of "admittance," defined as the vibration velocity divided by the driving force.[1] This measurement method is applicable to all string instruments.

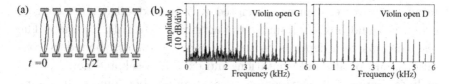

Figure 5.3: (a) The movement of a bowed string in one period. (b) The spectra of bowed open string G and D notes, showing a close correlation between maxima and minima in two spectra.

★ Example 5.1: What are the harmonics if you pluck a string at $\frac{1}{3}$ of the string length?

Answer: Since the harmonics 3, 6, 9 ... require a node at $\frac{1}{3}$ position of the string (see Fig. 2.9), a plucked string at the $\frac{1}{3}$ position *cannot* have a node at that position and thus the harmonics 3, 6, 9 ... cannot exist. The allowed harmonics are 1, 2, X, 4, 5, X, 7, 8, X, 10, ..., where X stands for missing harmonics.

---

[1]See Thomas Rossing (Editor), *The Science of String Instruments*, and Jim Woodhouse, The acoustics of the violin, *Rep. Prog. Phys.*, **77**, 115901 (2014); doi:10.1088/0034-4885/77/11/115901

B.   Resonance Box of String Instruments

There are many studies on the sound quality of the violin body (sound box) because of its tone quality. We discuss some results from violin sound box resonances as example in this section. These studies are applicable to the sound boxes of other string instruments. Violins were invented in Italy in early 1500s. Different parts of the violin are traditionally made from different types of wood: ebony and rosewood for the fingerboard, maple for the bridge, and spruce for the soundboard of the body. The soundboard amplifies the resonance of the strings and accounts for much of a final instrument's tonal quality, which depends on the resonant vibration of the violin body. In a review article, D. Murray Campbell discussed the complex nature of scientific measurements and subtle judgments by expert musicians on the evaluation of instruments.[2]

Resonance of the top and base plates can be analyzed by Chladni resonance pattern, named after Ernst Chladni (1756–1827). Chladni resonances are resonances associated with an extended object, e.g. a drum, a sound box, etc., under the action of external excitation. The resonance pattern depends on the frequency, geometry, boundary conditions, and position of excitation, and exhibits peak and nodal points or lines. Some of these resonances can produce big sounds, and some provide vibrational feeling to the performer. Different researchers have classified these resonances in slightly different ways. Bissinger *et al.*, classified resonances into six major vibration modes in two classes, reproduced in Fig. 5.4(a):[3]

(1) Cavity modes: (a) A0 mode at $f_{A0} \sim 280$ Hz, a Helmholtz-type mode characterized as a mass-plug oscillating under the influence of the cavity "spring," always a strong radiator, and (b) A1 mode, the first longitudinal mode at $\sim 1.7 f_{A0}$. These two modes are coupled.

[2]D. Murray Campbell, Evaluating musical instruments, *Physics Today*, **67**, 35 (2014); doi:10.1063/PT.3.2347

[3]See George Bissinger, *JASA*, **124**, 1764 (2008).

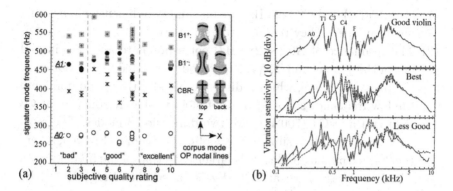

Figure 5.4: (a) Classifications of violin resonance modes by Bissinger into cavity and corpus modes. (b) Measured admittance of some violins and classification of violin resonance modes by Jansson into 1D, 2D, and Bridge–Hill (BH) modes around 2–3 kHz.

(2) Corpus modes: (a) CBR is the lowest frequency corpus mode with shear-like in-plane (IP) relative motion between top and back plates, where a ‡ out-of-plane (OP) nodal line pattern on top and back plates accompanied by out-of-phase f-hole volume flows. This is a relatively weak radiation. (b) The first corpus bending modes are B1$^-$ and B1$^+$. They both radiate strongly through the f-hole.

Erik Jansson at KTH in Stockholm carried out systematic studies on the sound quality of string instruments.[4] A small pendulum hammer carrying an accelerometer applies an impulse to the bridge, and a laser vibrometer measures the resulting bridge velocity. The sensitivity or input admittance is the velocity divided by the driving force. Figure 5.3(b) compares bridge-admittance vs frequency for modern violins, categorized as "good," "best" and "less good" violins. The vibration sensitivity measured at the bass side (left) and treble side (right) are marked with full and dashed lines. The measured admittance is reflected in the spectra shown in Fig. 5.2, where we find that the amplitude of the fundamental mode of the open G-note is lower than other harmonics. Jansson's classification of important violin resonances are as follows:

---

[4]See http://www.speech.kth.se/music/acviguit4/ (Erik Jansson)

(1) C1 (185 Hz) and C2 (405 Hz) are one and two-dimensional bending flexure modes. They radiate little sound.

(2) The resonance of A0 ($\sim$ 280 Hz) corresponds to air flows in and out of the f-hole; and T1 (475 Hz) mode is related to the motion of the top plate mode, which radiates sound effectively and dominate the low frequency sound spectra of most violins. Resonances at C3 (530 Hz) and C4 (700 Hz) are two-dimensional modes. The resonance is evidently visible in the violin spectrum of Fig. 5.5, where the 2nd harmonic amplitude is larger than that of the fundamental.

(3) The broad resonance at 2–3 kHz is called the Bridge–Hill (BH) mode.

Historically, the most famous violins are made by Stradivarius and Guarneri del Gesù. Many modern organologists carry out research to understand sciences of these outstanding violins. In 2009, Francis Schwarze, at the Swiss Federal Laboratories for Materials Testing and Research, found that the density of fungi-treated wood decreased slightly, but the speed of sound remained the same. The Swiss violin maker Michael Rhonheimer used fungi-treated wood to produce a Stradivari-quality violin.[5] A violin using woods treated by fungi for 9 months was "judged by 180 musicians to best a Stradivarius." However, Stradivarius and Guarneri del Gesù violins are worth millions of dollars. The historical value is similar to an art collection. In 2013, a violin played by the Titanic's bandleader as the ship sank sold at auction for more than $1.7M.

In 2012, the Univ. of Paris conducted experiments with 20 violinists to judge modern high-quality instruments vs Stradivari and Guarneri masterpieces. The astonishing result was that the "modern violin received highest appraisal from the experts, and Stradivari's was pronounced the worst." C. Fritz et al. carried out two separate comparative evaluations of "good new violins" and "old violins" of Stradivari or Guarneri, involving professional violinists and audiences (composed of violinists, violin makers, composers, and music critics) in an orchestra.[6]

---

[5]http://onlinelibrary.wiley.com/doi/10.1111/j.1469-8137.2008.02524.x/epdf; see also https://www.sciencedaily.com/releases/2009/09/090914111418.htm

[6]C. Fritz, J. Curtin, J. Poitevineau, and F.C. Tao, *PNAS*, **109**, 760 (2012); ibid, **114**, 5395 (2017).

Most evaluators considered "good modern-made violins" better at projecting and could not tell the new from the old apart when listening or playing blindfolded.

String instruments in the world come in many shapes and designs. They are played by picking, plucking, striking, and bowing. Their timbre depends on the way they are played. One can analyze these string instruments with similar methods as that of the violin discussed above. The design and style of playing of these specific string instruments reflect the characteristics of their local cultures.

II. Wind Instruments

Another important category of musical instruments is the wind instrument, classified into two categories: brasses, such as the trumpet and the tuba, and woodwinds, such as the clarinet and the saxophone. Figure 5.5 shows shapes and designs of some wind instruments. All wind instruments require a vibration source and a resonance chamber to produce resonance and amplify the musical sound.

Figure 5.5: Wind instruments divided into brass and woodwind with different shapes.

Sound Source and Bernoulli Principle in Wind Instruments

Sound sources are generated by the edge tone for flutes and by reeds (e.g. vibrating lips for brass instruments and thin sheets of vibrational medium for woodwind instruments). Reeds are classified into air/lip reeds for brass instruments, and single- and double-reeds for woodwind instruments.

The Bernoulli principle (see Appendix A) plays the key role for the function of reeds. When the speed of air passes through the opening of a reed at high speed, the pressure at the gap decreases. The tension of the reed forces the reed to close, decreasing the airflow until the pressure wave reflected from the end of the instrument forces open the reed for airflow to pass through again. This process sets up vibration of the reed and air pulses in sync with resonances of the instruments.

On the other hand, the air vibration of the flute instruments relies on the edge tone at the embouchure hole. As the air passes through a sharp edge, it creates turbulence. The airflow into the flute is maximum at frequencies of minimum impedance of the flute instrument. The resonances of flute instruments occur at the impedance minima.

Acoustic Impedance

The acoustic impedance inside an instrument is $Z_{Ai} = \frac{p}{U}$, where U is the volume flow rate and p is the sound pressure. It characterizes the performance of the instrument (see Eq. (2.21) in Chapter 2, Sec. VI). A smaller amount of airflow can set up a higher pressure-wave amplitude at the frequency where the impedance is maximum. When the impedance inside the instrument Z_{Ai} (given by $Z_{Ai} = \frac{\rho v}{A}$, where ρ is the density, v is the speed of the sound wave, and A is the cross-sectional area of the pipe) differs substantially from the impedance outside the instrument, wave reflections occur and set up interference for standing waves or resonances.

Resonances of Cylindrical Pipes

When there is substantial reflection of wave inside the cylindrical pipe, interference occur at standing wave frequencies. The resonance chamber of wind instruments is a column at various shapes for waves to achieve standing wave conditions. The body of the instrument can also produce "Chladni-plate-like" standing wave to enhance sound production at formant frequencies.

In an *open* pipe, the resonant frequencies are $f_n = nv/(2L)$ of Eq. (2.16), where v is the speed of the sound wave in the pipe, L is the length of the pipe, $n = 1$ is the fundamental, and $n = 2, 3, 4 \ldots$ are harmonics. In a *closed* cylindrical pipe, the resonant frequencies are $f_n = nv/(4L)$ of Eq. (2.17), where $n = 1$ is the fundamental, and $n = 3, 5, 7, \ldots$ are harmonics.

★ Example 5.2: The cylindrical air column of a B^b trumpet has a length of 140 cm. What are resonance frequencies of a closed end cylindrical pipe at air temperature 20°C?

Answer: At 20°C, the speed of sound is $v = 343$ m/s. The resonances of the closed-open cylindrical pipe air column is $f_n = n\frac{v}{4L} = n\frac{343}{4 \times 1.4} = 61, 184, 306, 429, \ldots$ Hz $(n = 1, 3, 5, \ldots)$.

A. Brass Instruments

Brass instruments include trumpet, French horn, trombone, and tuba. Figure 5.6(a) shows a schematic drawing of brass instruments that include mouthpiece, tapered matching mouthpiece, cylindrical section, and a bell. The player provides a pressure kPa, up to tens of kPa (the atmospheric pressure $p_0 \approx 101$ kPa), to the instrument. The acoustic energy radiates out mainly from the bell, with some energy lost in friction. One can change the resonance frequencies by modifying the "effective length" of the instrument via a slide or valve, as shown in Fig. 5.6(b).

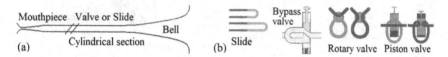

Figure 5.6: (a) Schematic drawing of a wind instrument made of cylindrical pipe with added mouthpiece and bell. Bell alters the sound spectrum and brilliance, and the tapered mouthpiece can provide ease of playing with impedance matching. (b) Schematic drawing of slides or valves that change the effective length of the instruments.

Flare, Bell and Mouthpiece

A flare is a gradual widening of the instrument tube, resembling conic shaped. A bell is a rapid widening at the end of the instrument (see Fig. 5.6). The bell is normally shaped in Bessel-horn geometry with $D = B/(y - y_0)^m$, where D is the diameter, B and y_0 are parameters chosen to provide proper diameters at the small and large ends, and m can range from 0.5 to 0.65 for a typical trumpet, and 0.7 to 0.9 for

French horn. The mode frequencies, including the bell, are

$$f_n = \left[(2n-1) + 0.637\sqrt{m(m+1)}\right]\frac{v}{4(L_0 + y_0)}. \qquad (5.1)$$

Here L_0 is the length of the cylindrical section. Typically, y_0 is about $2 \sim 3$ cm. If we choose $m = 1$ for the Bessel geometry, the mode frequencies become $f_n \approx nv/(2L)$ for $(n = 1, 2, 3, \dots)$, where L is the effective length of the brass instrument, and we approximate the coefficient 0.637 with $1/\sqrt{2}$. In summary, resonance frequencies of brass instruments without flare and with flare are respectively $f = \frac{v}{2L} \times \left(\frac{1}{2}, \frac{3}{2}, \frac{5}{2}, \dots\right)$ and $f = \frac{v}{2L} \times (1, 2, 3, \dots)$, where L is the "effective" length. The bell makes the spectrum open-tube-like. Figure 5.7(a) shows the acoustic impedance of a trumpet with and without flare or bell, with and without mouthpiece.[7]

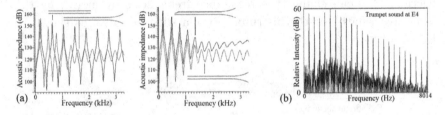

Figure 5.7: (a) Comparison of acoustic impedance of brass instruments with/without the bell and mouthpiece. (b) The spectrum of a trumpet sound playing E_4 (330 Hz).

With a flare or a bell, the higher frequency waves can efficiently radiate outwards, producing the characteristic bright and loud sound of brass instruments. However, smaller high frequency reflections produce weaker standing waves. Weak resonances give a rather flexible notes controlled mainly by the vibrational frequency of the lips. The high frequency radiation from brass instruments is directional, i.e. substantially louder when the bell is pointed at you. The mouthpiece provides

---

[7]See http://newt.phys.unsw.edu.au/jw/brassacoustics.html for details. This website provides outstanding resources for all wind instruments. Figure 5.5(a) is adopted from this website.

an impedance matching from the mouth to the cylindrical tube so that it is easier to play, indicated by higher impedance at the resonance peaks in Fig. 5.7(a). The mouthpiece strengthens some of the resonances, lowers the frequency of the very highest resonances, and most importantly, increases the high frequency harmonics. Figure 5.7(b) shows the spectrum of a trumpet playing the E_4 at 330 Hz.

★ Example 5.3: What are the resonance frequencies of a brass instrument with a Bessel bell? Is it a closed pipe or open pipe?

Answer: As discussed in Eq. (5.1), with a Bessel bell, the standing wave frequencies of a trumpet can be approximated by $f_n = n\frac{v}{2L}$ for $(n = 1, 2, 3, \ldots)$. It is an open pipe.

Valves or Slides

Brass instruments use valves or slides to change the effective length and fill in notes of the chromatic scale that lie between modes of the open instrument (see Fig. 5.4). Based on the equal tempered scale, one can increase the acoustic length by $[2^{2/12} - 1] = 12.2\%$ to lower the pitch a whole tone or increase the length by $[2^{1/12} - 1] = 5.89\%$ to lower the pitch by a semitone. For example, the first valve on a trumpet lowers the pitch by a whole step; the second by a half step; and the third by three half steps. Trumpets have a slide on the pipe for easy tuning. This can also be achieved by a valve on the trumpet, as it is with the tuba. A slide is also used on the trombone.

★ Example 5.4: The effective length of a trombone is $L = 2.75$ m in "first" position, with the slide fully retracted. To lower the pitch by one whole tone (by 12.2% in moving to position 3), what is the length of slide change in order to change the note by one whole tone?

Answer: The length change of the trombone is $\Delta L = 12.2\% \times L = 0.34$ m. Since the total length of the air column in the trombone is twice the slide length, the slide needs to change by 0.17 m.

★ Example 5.5: Pressing the first valve of a trumpet or tuba increases the acoustic length by 12.2% and lowers the pitch a whole tone. The "effective" length of an instrument is the length of an idealized cylinder (or cone) that gives the appropriate frequency for its fundamental mode. If the "effective" length of the trumpet is 1.40 m, what is the length of tubing that should be added to produce this amount of pitch change?

Answer: Total length change is $\Delta L = 12.2\% \times L = 0.17$ m.

Mute

The bell, apart from its effect on tuning, radiates high harmonics easily. High harmonics in the output sound make brass instruments sound bright and "brassy." The mute in the bell reduces this radiating effectiveness for most frequencies, and the overall timbre becomes less bright. However, there is usually a band of frequencies that the mute transmits well, giving rise to a "formant" in the output sound — a broad band of strong harmonics. For trumpet and cornet mutes, standard mutes produce strong formants at various frequencies in the 1–3 kHz range.

B.   Woodwind Instruments

Woodwinds include the single-reed instruments (clarinets and saxophones), the double reed instruments (oboes, bassoons, English-horns), and the "edge-tone" instruments (flutes, piccolos, and recorders). Modern woodwinds may be made of wood, metal, or plastic.

Reed and Edge Tone

The sound sources of woodwind instruments are reeds and the edge tone for flutes. Figure 5.8 shows some schematic drawing of a single reed; double reed; and turbulent airflow in edge tone for flute instruments.

Figure 5.8: Schematic drawing of a single reed, double reed, and edge tone (sound source of flute instruments).

The working principle of a reed is the Bernoulli principle. Figure 5.9 schematically shows air-volume flowrate vs the pressure difference between the mouth and the mouthpiece in the reed. Typically, the volume flow rate is 10–100 [cm^3/s] and the pressure difference is tens of [kPa]. As the pressure difference increases, the airflow rate increases until it reaches a maximum. After that, the airflow decreases as the pressure increase until the reed is shut-off due to the tension of the reed. A reed-valve can sustain periodic oscillation in an air column only when the pressure is in the negative-slope region of the flowrate-vs-pressure diagram, i.e. with impedance $Z \equiv \Delta p/U < 0$. In the negative-slope regime, the pressure wave, reflected from the end of the instrument, helps to force the reed open again for airflow to pass, i.e. the reed position is the antinode of pressure standing wave. This process set up the reed's vibration. The frequency of the airflow pulse is the same as the resonant frequency of the air column of the musical instrument.

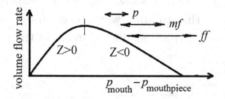

Figure 5.9: The working principle of reeds in woodwind instruments is the Bernoulli principle. Stable operation of the reed is on the downward side of the airflow vs pressure region, with negative impedance $Z < 0$. Players can control the loudness and harmonic content by varying the pressure difference between the mouth and the mouthpiece.

Figure 5.9 also shows possible playing scenarios by varying the pressure differences. If one plays softly, the pressure wave is nearly sinusoidal. Played louder, the sound wave becomes nonlinear. At a very loud sound level, the airflow and the pressure wave may be clipped, and its waveform is sharp. Its spectrum will contain many high harmonics (Fourier theorem). The hardness of a reed can also affect the spectrum. A hard reed will produce a louder sound at high frequencies.

For the edge tone musical instruments, experiments and analyses show that the frequency depends on the air speed passing an edge, while the resonances of the pipe provide stable sound at the resonance frequencies.[8] The resonance location in flute instruments has minimum impedance so that airflow can easily enter the flute column to set up standing waves.

Bore Type

Woodwinds use the cylindrical or conical bores, where flute and clarinet have cylindrical bore. Most other woodwinds (oboe, English horn, bassoon, saxophone, etc.) are essentially conical. The cone angles of oboe and bassoon are small ($1.4°$ and $0.8°$), whereas those of the saxophone are large ($3°$ and $4°$). The recorder is a flute type instrument having a reverse conical bore and a fixed wind pathway.

Resonances of a closed conic cylindrical pipe have essentially the same frequency spectrum as an open pipe of the same length, even if the cone is truncated. Figure 5.10 shows the calculations by Ayers, Eliason and Mahgerefteh on the standing wave spectrum of various geometric conical pipes, where r_2/r_1 stands for the ratio of the radii at both ends

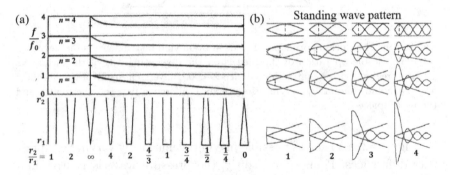

Figure 5.10: (a) Frequency ratio of standing wave harmonics vs geometry of the pipe, where $f_0 = v/2L$ with the length of the cylindrical pipe being L. The ratio r_2/r_1 stands for the ratio of the radii of the ends. For an open uniform cylindrical pipe, the ratio is 1; a conic geometry has a ratio of infinity (∞). For a uniform closed cylindrical pipe, the ratio is 1. In this case, the resonances occur at $\frac{v}{2L}(\frac{1}{2}, \frac{3}{2}, \frac{5}{2}, \ldots)$. (b) Standing wave modes ($n = 1, 2, 3, 4$) in open cylindrical pipe with $r_2/r_1 = 1$ to ∞.

---

[8]See J.W. Coltman, *JASA*, **60**, 725 (1976).

of the cone, and their standing wave patterns.[9] The cylindrical open pipe corresponds to $r_2/r_1 = 1$. Flutes are cylindrical open pipe. As one end getting smaller, the standing wave spectrum does not change, up to $r_2/r_1 = \infty$, which is a nearly a closed cone-shaped pipe. The standing wave frequencies are $f_n = n\frac{v}{2L}$, with $n = 1, 2, 3 \dots$ The conical woodwinds use the closed cone-pipe geometry with $r_2/r_1 = \infty$. The reed of the mouthpiece is the closed conical end.

If one widens the closed end, the ratio r_2/r_1 decreases, and the resonance frequency decreases as well. When it reaches $r_2/r_1 = 1$, the pipe becomes a closed cylindrical pipe with the closed pipe spectrum and has only the odd harmonic resonances at $f = \frac{v}{2L}\left(\frac{1}{2}, \frac{3}{2}, \frac{5}{2}, \dots\right)$. The clarinet is an example of a closed cylindrical pipe.

Tone Holes

All woodwinds use tone holes to change the effective length in order to produce the proper music notes. The effective length depends on the size of the tone hole, as shown in Fig. 5.11(a). When the hole size matches the cylinder bore, the effectively length is the distance from the tone-hole to the playing end. Expert musicians may partially cover the tone hole to produce slighted varied tones. If the open holes are regularly spaced to form a "tone-hole-lattice," it acts as a filter that transmits high frequencies but reflects low frequencies. The "cutoff frequency of the lattice" is an important factor in determining the timbre of the instrument, shown in Fig. 5.11(b).[10]

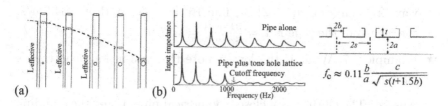

Figure 5.11: (a) Effective length vs tone-hole size of wind instruments. (b) Tone-hole-lattice can also affect high frequency spectrum of flute instruments.

---

[9]R.D. Ayers, L.J. Eliason, and D. Mahgerefteh, Conical bore in musical acoustics, *Am. J. Phys.*, **53**, 528 (1985), adopted in Fig. 5.8.

[10]See Ref. [23] and J. Wolfe and J. Smith, *JASA*, **114**, 2263 (2003).

Register Holes

A register hole or vent is a special tone hole that forces a nodal position of the standing wave. When the register opens, the standing wave at this location must be a "node" (see Fig. 2.11), and it forces the instrument to resonate at the next harmonic. Figure 5.12 shows a schematic drawing of a register of a clarinet. Ideally, the clarinet is a closed pipe, and the register hole should be located at $\frac{1}{3}$ of the distance between the reed and the tone hole. The nodal point at register-hole position forces the instrument to resonate at the next higher harmonic. For conical woodwinds and flutes, the position of the register hole should be located at the mid-point between the tone hole and reed position for an open pipe so that it forces a resonance at the next harmonic.

Tone hole
Register key closed
Fundamental Mode

Tone hole
Register key open
Third Harmonic Mode

Figure 5.12: Because the clarinet is a closed cylindrical pipe, the standing wave will have an antinode at the mouth-piece (see left plot). By opening the register hole, the location becomes a pressure node of a standing wave. The allowed standing wave is the third harmonic.

★ Example 5.6: What is the effective length of a hyperbass flute to play C0 at 16 Hz?

Answer: The flute is an open cylindrical pipe. Using the standing wave frequency of Eq. (3.2), we find 16 Hz = $\frac{v}{2L}$. With $v = 343$ m/s, the length of the pipe is $L = \frac{343\,\text{m/s}}{2 \times 16\,\text{m}} = 10.7$ m.

★ Example 5.7: What is the lowest note that a 3-m long contrabass clarinet?

Answer: The clarinet is a closed cylindrical pipe. Using the standing wave frequency of Eq. (3.3), we find the lowest frequency of the 3-m clarinet is $f = \frac{v}{4L} = \frac{343[\text{m/s}]}{4 \times 3[\text{m}]} = 28.6$ Hz.

★ Example 5.8: What is the function of register key?

Answer: The opening of the register key forces a pressure nodal point at the register location to cause resonance at a higher harmonic, as shown in Fig. 5.12.

★ Example 5.9: Where is the optimal location of the register keys for woodwind instruments?

Answer: Figures 2.11 and 5.10 show standing wave patterns. Since the clarinet is a closed cylindrical pipe, the register hole is ideally placed about $\frac{1}{3}$ the distance between the reed tip and the first open hole. For open-cylinder-pipe or conic-shaped-type musical instruments, e.g. saxophones and oboes, the register hole is ideally placed at $\frac{1}{2}$ the distance between the cone apex and the first open hole for the next harmonic.

C.   Other Musical Instruments

There are percussion musical instruments such as drum, cymbal, gong, bell, rattle, xylophone, glockenspiel, steel-pan or space-pan (drum), etc. played by striking one object against another. These instruments have varying shapes and sizes, as shown in Fig. 5.13. The Chladni-pattern-like resonances of the percussion instruments are not integer multiples of a fundamental frequency, and their temporal wave pattern is not periodic. Even though the sound waves of any percussion instruments are not periodic, they can provide rhythm and attention. They are integral part of orchestra music, rhythmic music, military march music,

Figure 5.13: Various types of percussion instruments. Some pitched percussion instruments are designed to diatonic scale.

modern pop, dance music and rock music. The sound produced by percussion instruments, including gunpowder, can dramatically enrich music performance. Composers employ percussion instruments to control the rhythm and develop and generate dramatic effects in music.

The distinctly rich and crispy sound of the "tuned percussion instruments," such as marimba, steel pan, gong, xylophone, glockenspiel, chimes, etc., have been popular in dance music. Musicians have composed music scores specifically for these tuned percussion instruments. They have also become popular in orchestra around the world.

III.   Spectrum and Timbre of Musical Instruments

The harmonic content of musical instruments depends on the resonances and structure of the instrument body and on the proficiency of the performer. Since the temporal wave pattern of string and wind instruments is periodic, its harmonic content comprises integer multiples of the fundamental, in contrast to that of percussion instruments. Figure 5.14 shows spectra of a clarinet, a trumpet, and a gong. Note that the spectrum of the clarinet clearly shows dominant odd harmonics at low frequencies, while the trumpet shows brilliant high harmonic content. The waveforms of both instruments show periodic structure. The wave pat-

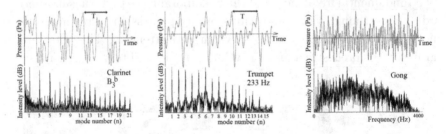

Figure 5.14: Comparison of waveforms and spectra of a clarinet, a trumpet, and a gong. The clarinet is clearly a closed cylindrical tube with missing even harmonics (except high frequency harmonics). The spectrum of the trumpet is equivalent to an open pipe. The period T of the waveform is the inverse of the fundamental frequency. The formants of each instrument are visible from their spectra and wave pattern. The gong is a percussion instrument. Its resonance is the same as the Chladni plate. The peaks in frequencies are not harmonics of each other.

tern of the gong does not have periodic structure, and its spectrum is not harmonic. The inharmonicity of the gong can draw attention to an important event during a music performance. To the extreme, composers may include gunpowder and fireworks into a music score for a dramatic finish to a performance.

There are also pitched (or tuned) percussion instruments made of wood, marble, or metal with varying shape and sizes of plates with additional resonant chambers. The standing wave of these extended objects may produce integer multiples of a pitch frequency and other non-pitched resonant frequencies. Figure 5.15 shows the waveforms and spectra of xylophone and spacepan. There are many low frequency resonances besides the harmonics of the pitch frequency. Since the standing wave depends on the boundary condition and the excitation position of the percussion instruments, the pitch and timbre of a tuned percussion musical instrument depend on the mallet or hand striking position. The characteristic crisp timbre of pitched percussion instruments arises mainly from their pulse envelopes and their characteristic spectra.

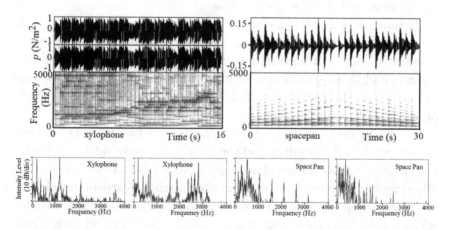

Figure 5.15: (Top) The waveforms and spectra of the xylophone and space-pan. (Bottom) The spectrum of a single note of xylophone and space-pan. The high frequency contents in the spectrum of xylophone provides the "ringing bell characteristics." High frequency components in the spectra of steelpan or spacepan decay faster than that of low frequency.

IV.   Summary and Other Musical Instruments

All cultures have developed their own musical instruments. The earliest known string instruments began being developed before 500 BC. A recently unearthed bone flute, found at Hohle Fels Cave in Germany, is at least 42,000 years old.[11] All musical instruments are made of a sound source and a resonant body to produce enhanced musical tones. Vibrating media include strings, reeds, an edge tone, a membrane or drumhead, wood bars, and pipes.

Musical instruments rely on standing waves to produce loud and steady sound. The sizes of all musical instruments are inversely proportional to the lowest frequency intended. String instruments also use strings of different gauges (sizes) or wrap-wire to change the mass per unit length and therefore change the transverse wave speed on the string. This changes the standing wave frequency without changing the size of the instruments. Modern musical instruments are normally designed or tuned to produce equal temperament tuning with the octave as the basic unit in most musical scales.

Resonance bodies include sound boxes in string instruments, pipes of various geometries, drum boxes, and various geometries of percussion instruments. All resonance bodies produce Chladni-plate-like resonances and Helmholtz air-spring-like resonance. These resonances are "formants" of the instrument. The formant of the instrument depends on the size of the instrument.

We normally classify wind instruments into two main categories: brasses, such as the trumpet and the tuba, and woodwinds, such as the clarinet and the saxophone. The sound sources of wind instruments are vibrating lips for brass instruments, an oscillating air stream (or air jet) for flutes, and a vibrating reed for woodwinds. The Bernoulli principle, which states that faster airflow causes lower pressure and vice versa, provides the basic mechanism of the periodic vibration of the sound source. The combination of the Bernoulli principle and the standing wave can produce a steady sound wave vibration at the resonance frequency of the sound box.

---

[11]See Thomas Higham *et al.*, *J. of Human Evolution*, **62**, 664 (2012); http://www.sciencedirect.com/science/article/pii/S0047248412000425

For wind instruments, the sound box is generally a cylindrical pipe. The cylindrical air column is generally either a closed or an open pipe. The closed pipe has standing wave frequencies at the odd integer harmonics of the fundamental, i.e. f_0, $3f_0$, $5f_0$..., which is less desirable in tuning. Human ingenuity modified these closed pipes by introducing flares or by changing the pipe into a conic cylinder. In this way, the standing wave frequency becomes open-pipe-like. However, the air column of a clarinet is a closed cylinder to provide its special timbre. Since the edge tone position of a flute is not at the closed end of the instrument, the air column of flute is an open cylinder. Some cultures even use spherical balls, melon shells, etc., for the resonance box to produce specialized music notes.

Brass instruments use valves or slides to change its effective lengths and fill in notes of the chromatic scale that lie between modes in the open instrument. Meanwhile, woodwind instruments employ tone holes to change the effective length of the resonance chamber. The chromatic scale notes normally conform to equal temperament tuning. Woodwind instruments use register keys to force a pressure "node" at the register key location and produce a higher harmonic standing wave.

Keyboard instruments, such as Piano, harpsichord, and pipe organs, have similar resonance structures as string and wind instruments. The piano, invented by Bartolomeo Christifori in Florence, has become a very important instrument in the orchestra. A typical concert grand piano has 243 strings, varying in length from about 2 m to 5 cm. There are 88 keys (A0-C8). The soundboard is made of spruce of about 1 cm thick. The soundboard, serving as a resonating box, can enhance the power of string instruments. The timbre of piano, played by striking the string with a hammer, resemble that of percussion instruments.

Percussion instruments employ bars, membranes, plates, and even gunpowder. Even though the sound waves of any percussion instruments are not periodic, they can produce beautiful sounds. They are an integral part of orchestra music, rhythmic music, military march music, and modern pop and rock music. The sound produced by percussion instruments, including gunpowder, can dramatically enrich a musical performance. The distinctly rich sound of the "tuned percussion instruments," such as the marimba, steel pan, gong, xylophone, glockenspiel, chimes, etc., has been popular in dance music. The spectra of these tuned percussion instruments may have essentially a single peak or harmonic structure with a broader bandwidth depending on their design. Music scores composed specifically for these tuned percussion instruments have become popular in orchestras around the world.

Advances in computer technology can enhance and modify the sound qualities of all musical instruments, including percussion. Modern electronic musical devices can emulate sounds from various musical instruments, including the human voice, or generate sounds beyond the traditional musical instruments. Synthesizers have enriched modern pop music and created new musical sound sources, including the electric guitar, electric violin, electronic keyboard, electronic enhanced piano, and electronic percussion beat generator. They can even modify voices to produce special effects in performances. In reality, music is very experiential. People tend to stick to what they are familiar with. Nevertheless, some people are excited about music purely generated by electronic synthesizers on speakers (electroacoustic music).

V.   Homework 05

1. All musical instruments are made of a vibrating source and a resonator to amplify the sound. The vibrating source of violin is a _____, a drum is a _____, a marimba is a _____, an organ is a _____, and a wind instrument is _____.

2. _____, the basic unit in most musical scales, refers to frequencies at the ratio of 2:1.

3. A string is plucked at $\frac{2}{3}$ location of its length. The allowed harmonics are _____, _____, _____, _____, _____ ... and a string is plucked at $\frac{2}{5}$ of its length location, the allowed harmonics are _____, _____, _____, _____, _____, ...

4. The standing wave frequencies of a conic instrument behaves like (open/closed) (choose one) cylindrical pipe.

5. Wind instruments are classified into two categories, _____
   such as the trumpet and the tuba; and _____ such as
   the clarinet and the saxophone.

6. The sound source of a wind instrument comes from an oscillating
   _____ or from a vibrating _____.

7. Brass instruments use _____ or _____ to
   change the lengths and fill in notes of the chromatic scale that
   lie between modes in the open instrument.

8. (choose one) The air column of a clarinet is a (open/closed) cylinder,
   the air column of a flute is a (open/closed) cylinder, and the air
   column of an oboe is a (conic/straight) cylinder.

9. A steel violin string vibrates at a fundamental frequency of 440 Hz.
   The length of the string is 0.33 m and the mass of the string
   is 0.24×10^{-3} kg. The wavelength of the wave on the string is
   _____ m, and the tension is _____ N. What
   is the wavelength of the A4 sound wave in room at 20°C?
   _____ m.

10. The piano string W27 has a mass per unit length $\mu = 0.0184$ kg/m.
    The length of the A0 key of a piano string is 1.22 m, what is the
    wavelength of the fundamental mode on the string? The tension
    on the string is _____ N. The wavelength of the A0 note
    in the room of 20°C is _____ m. What will be the wave-
    length of A0 note playing in a room 10°C? _____ m.

11. If an organ pipe was tuned to 440 Hz at 20°C, the frequency at
    30°C = _____ Hz.

12. If an organ pipe was tuned to 440 Hz at 20°C, the frequency at
    10°C = _____ Hz.

13. The air column (without a flare) of a B-flat trumpet has a length
    of 1.40 m. Using the model of a closed cylindrical pipe, the first

three allowed resonances at 20°C are $f_1 =$ _____ Hz,
$f_3 =$ _____ Hz, and $f_5 =$ _____ Hz.

14. The _____ dependence of the radiation efficiency of the _____ has an important impact on the spectrum of radiated sound for a brass instrument.

15. Pressing the first valve of a trumpet or tuba increases the acoustic length by 12.2% and lowers the pitch a whole tone. The "effective" length of an instrument is the length of an idealized cylinder (or cone) that gives the appropriate frequency for its fundamental mode. If the "effective" length of the trumpet is 1.40 m, _____ cm of tubing should be added to produce this amount of pitch change.

16. The effective length of a trombone is 2.75 m, in "first" position with the slide fully retracted. To lower the pitch by one whole tone (by 12% in moving to position 3), the length must increase by _____ cm, which means that the slide must move by _____ cm.

17. The lowest resonance frequency of a clarinet (a closed cylindrical pipe) is about 67 cm long is _____ Hz at 20°C air temperature.

18. The lowest note of a 32-inch open organ pipe is _____ Hz in the room at 20°C. The wavelength of the sound wave in the organ pipe is _____ m. The wavelength of the sound wave in the room is _____ m.

19. The lowest note of a 32-inch open organ pipe is _____ Hz in the room at 10°C. The wavelength of the sound wave in the organ pipe is _____ m. The wavelength of the sound wave in the room is _____ m.

20. (choose one) A register hole or vent is designed to force a "pressure" (node/antinode) at the open vent location so that the resonance of the air column oscillates at a strong higher harmonic.

Chapter 6

Human Voice, Speech, Singing, Analysis and Synthesis

Musical instruments are composed of a sound source and a resonator for the amplification and modulation of the desired sound. Similarly, the human voice is produced by a sound source and a resonant chamber that amplifies and modulates the sound for articulation. The sound source is the vibration of vocal folds or from friction and restriction of airflow at our lips, teeth and vocal tract. The resonant chamber is our vocal tract and nasal cavity. According to a linguistic survey, there are about 7,000 living languages in the world. The "phoneme" is the basic unit of speech sound in all spoken languages. We divide the phonemes (articulation of speech sounds) into vowels and consonants. In English, speech scientists list 12-21 vowel sounds and classify consonants into plosive (t, p, k, b, d, g, ...), fricative (f, s, sh, ... and h), nasal (m, n, ng), liquid (r, l), and semi-vowel or glide vowel (w, y). The human auditory system has a maximum sensitivity in the frequency range of 1,000 to 4,000 Hz (see Fig. 3.4 in Chapter 3), which is also the range of resonance frequencies (formants) of our vocal tract. The human voice and ears are well-matched organs. This chapter analyzes the basic physics of human speech and singing.

I. Speech Production

Speech production is divided into sound source and articulation. Human organs involved in speech production include the lung, the larynx, and the vocal tract.

A.   Voice Source

The lung serves as the air source (engine). As air flows through the vocal box, the vocal cords modulate airflow into periodic air pulses that have rich frequency content. The vocal tract can dynamically change shape to articulate the speech sound. Figure 6.1 shows a schematic drawing of the human vocal system and an expanded anatomical drawing of the larynx, which includes vocal folds and the glottis (the opening between the vocal folds). The length of the vocal folds is about 17–25 mm in an adult male and 12.5–17.5 mm in an adult female, with a volume of 60 cm^3. The vocal folds open and close to modulate airflow per the Bernoulli principle. As air moves faster through the glottis (vocal fold opening), the pressure at the glottis decreases, and the tension of the vocal folds forces the opening to close, until the sub-glottal pressure builds up and forces the vocal folds to open again. The natural vibrational period depends on the mass and elasticity of the vocal folds, and on personal physical condition. The rapid airflow produces a buzzing sound.

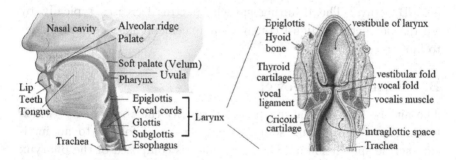

Figure 6.1: Schematic drawing of human vocal system and the expanded view of the larynx.

Figure 6.2 shows stroboscopic photos of successive vocal folds vibration in one cycle, a schematic drawing of a periodic air-pulses waveform of period 8 ms, and a corresponding sound spectrum at harmonics of 125 Hz. The pitch is the vocal fold vibration frequency. It shows up as the spacing between vertical lines in the frequency spectrum. Typically, the pitch is about 100 Hz for men, 200 Hz for women, and 300 Hz for children in normal speech. In singing, a bass singer can produce a pitch at 60–70 Hz, and soprano up to 1,170 Hz. The typical shape of an

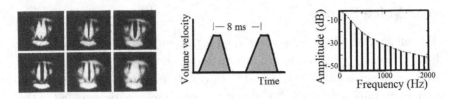

Figure 6.2: Stroboscopic photos showing successive vocal fold vibration in 1 cycle; a *schematic* example of air pulses with an 8 ms period; and a spectrum of 125 Hz harmonics.

air pulse has a triangular-like waveform that has the power spectrum reducing by 12 dB per octave (see Table 4.1), agreeing with measurements (see Ref. [30] for details). The output intensity of a speech sound is

$$(\text{Output intensity}) = (\text{source}) \times (\text{filter function}) \times (\text{radiation efficiency}). \tag{6.1}$$

The mouth may radiate more efficiently at higher frequency with +6 dB/octave. The resulting speech spectra decreases typically by −6 dB per octave. High-frequency enhancement makes sound louder to human ears.

B.   Formant and Vowel

The airflow from the larynx passes through "the vocal tract," which transforms the "buzzing sound" from the vocal folds into an intricate subtle speech sound. The vocal tract is composed of the pharynx (throat), nasal cavity, and oral cavity. The oral cavity includes a flexible tongue, lips, cheeks, and teeth. All spoken human languages require dynamic modulation of the buzzing sound source by the vocal tract. Altering the shape of the vocal tract can produce vocal cavity resonances, transforming buzzing sounds into speech sounds. "Formants" are spectral peaks of a speech sound or resonances of the vocal tract. Each individual human voice depends on the shape and size of the vocal cords, vocal tract, lung capacity, health condition, and habit and manner of forming words. Skilled imitators can mimic the manner of forming words in artistic performances.

★ Example 6.1: Find the standing wave modes of a 17 cm closed cylindrical pipe at 20°C.

Answer: The resonances for the closed cylindrical pipe is $f_n = n\left(\frac{v}{4L}\right) = n\left(\frac{343}{4 \times 0.17}\right) = n \times 500$ Hz, $n = 1, 3, 5, \ldots$, where we use the speed of sound wave $v = 343$ m/s at 20°C.

Figure 6.3(a) shows the response of the closed cylindrical pipe to sine waves at varying frequencies (top), as well as the response to a "single" sawtooth wave at 100 Hz (bottom) for modeling the vocal tract as a closed cylindrical pipe of 17 cm.

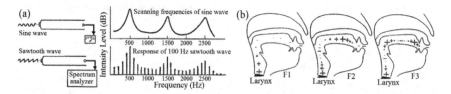

Figure 6.3: (a) Schematic drawing of the transfer function of a sine wave and a sawtooth wave input into a 17-cm closed tube. (b) Schematic drawing of 3 standing waves (formants) of a vocal tract. The $+/-$ signs signify antinode/node positions.

When a sinusoidal wave drives a closed cylindrical air column of 17 cm at air temperature 20°C, each sinusoidal wave will produce a unique response amplitude at that frequency. The maximum response occurs at resonance frequencies 500 Hz, 1500 Hz, 2500 Hz, etc. The top plot of Fig. 6.3(a) shows the response by scanning frequencies of the sinusoidal waves. Each mode has its own resonance width or Q-factor. On the other hand, a single sawtooth wave at 100 Hz contains all harmonics of the 100 Hz. The response will excite all frequencies at harmonics of 100 Hz, following the envelope of standing wave modes, as shown in the bottom plot of Fig. 6.3(a). Figure 6.3(b) schematically shows the standing wave pattern of the closed pipe of our vocal tract. The $+$ sign is the pressure antinode, and the $-$ sign is the pressure node.

The vocal track is much more complicated than a cylindrical pipe. Figure 6.4(a) shows a schematic plot of two speech spectra of vowels /ah/ and /e/ at the fundamental pitch of 150 Hz. Figure 6.4(b) shows a schematic drawing of various vowels and the associated shapes of the vocal track, derived from X-ray imaging of spoken vowels. Figure 6.5(a) lists the first three formants of some vowels for men, women, and children. Figure 6.5(b) shows the frequency range of the first two formants of some vowels.

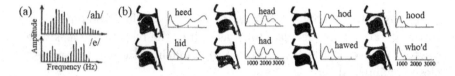

Figure 6.4: (a) Examples of speech spectra for the vowels /ah/ and /e/ with the fundamental pitch of 150 Hz. (b) Examples of vocal track shapes for various vowels.

(a)

Vowel		Men			Women			Children		
		F1	F2	F3	F1	F2	F3	F1	F2	F3
beat	/ee/	270	2300	3000	300	2800	3300	370	3200	3700
bit	/i/	400	2000	2550	430	2500	3100	530	2750	3600
bet	/e/	530	1850	2500	600	2350	3000	700	2600	3550
bat	/ae/	660	1700	2400	860	2050	2850	1000	2300	3300
part	/ah/	730	1100	2450	850	1200	2800	1030	1350	3200
Pot	/aw/	570	850	2400	590	900	2700	680	1050	3200
boot	/u/	440	1000	2250	470	1150	2700	560	1400	3300
book	/oo/	300	850	2250	370	950	2650	430	1150	3250
but	/u/	640	1200	2400	760	1400	2800	850	1600	3350
pert	/er/	490	1350	1700	500	1650	1950	560	1650	2150

(b)

Figure 6.5: (a) A Table for three formants of common International Phonetic Alphabet (IPA) vowels; (b) Frequency range of the first two formants of various vowels. The dashed lines mark the human singing range of about 80–1,100 Hz.

★ Example 6.2: Identify the pitch of three spectra of the vowel /ε/ sound below.

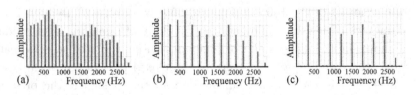

Answer: The pitch is the lowest frequency or the spacing of the spectral lines. Their pitches are 100, 200, and 300 Hz, respectively.

C.   Regional Dialect of Speech

All vowels can be distinguished by their formants, or resonances of the vocal track. All languages can evolve into regional dialects that sometimes may be difficult for outside people to communicate. For example, some regional dialects of a language may evolve into completely different phoenetics not understandable to people in different regions, even though the written forms are mutually comprehensible. Some dialects have completely different vowel sounds.

The formant forming of a child is an interesting research topics for linguists. Linguists found that the formants of a person depend on whether the person has learned the language before or after the age 6. Figure 6.6 shows a survey of the first two formants in English vowels for people in the US and Australia.[1] The similarity of these vowel maps implies that people can easily communicate with each other.

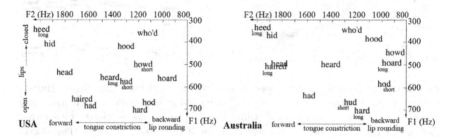

Figure 6.6: Comparison of the relative frequency correlation of the first two formants of most English vowels in the United States and Australia. People in different regions may form different formants and different dialects. This is true of all languages.

D.   Consonants

In English, linguists classify consonants into plosive (t, p, k, b, d, g ...), fricative (f, s, sh ... and h), nasal (m, n, ng), liquid (r, l), and semi-vowel or glide vowel (w, y). The plosive consonants result from constricting or blocking the flow of air and suddenly releasing it in milliseconds at a place in the vocal tract, e.g. the lips, alveolus and velum (see Fig. 6.1).

---

[1] http://newt.phys.unsw.edu.au/jw/voice.html

In 1952, Haskins Laboratories used the "pattern-playback device," which can produce sound from a given spectral pattern, to study the recognition of plosive consonants. Cooper *et al.* presented listeners with a noise burst of 15 ms at varying frequency ranges, and a subsequent 2-formant vowel sound for the recognition of consonants /t/, /p/ and /k/. Figure 6.7(a) shows that high frequency bursts ($\geq$ 2500 Hz) produce /t/ for all vowels, but lower frequency bursts near formant frequencies give /k/, shown as the shark-shaped distribution. Low- and medium-frequency bursts not too near the formant frequency generate /p/ sound.[2]

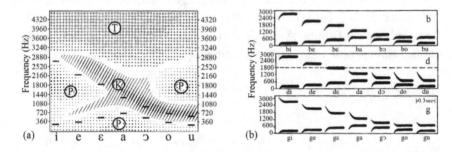

Figure 6.7: (a) Results from Cooper *et al.* on the recognition of the plosive (burst) frequency consonants /t/, /k/, and /p/, with 2-formant vowels. High-frequency burst produce the consonant /t/, while low-frequency plosive bursts give either /k/ or /p/ depending on the relative position of the burst frequency. The shark-shaped burst sound gives consonant /k/. (b) Frequency transition to formants in Cooper's experiments that led to identifying voiced plosive consonants (/b/, /d/, and /g/).

Consonant recognition depends also on the transition into the formants of the vowel sound. The frequency transition may be associated with the consonant articulatory places. It appears that the voice plosive consonants correspond to the frequency transition of the first formant, while the voiceless consonants do not have frequency transition of the

[2]See http://en.wikipedia.org/wiki/Haskins_Laboratories; F.S. Cooper, P.C. Delattre, A.M. Liberman, J.M. Borst, and L.J. Gerstman, *JASA*, **24**, 597 (1952); A.M. Liberman, P.C. Delattre, F.S. Cooper, and L.J. Gerstman, *Psychological Monographs: General and Applied*, **68**, 1 (1954), http://dx.doi.org/10.1037/h0093673; P.C. Delattre, A.M. Liberman, and F.S. Cooper, *JASA*, **27**, 769 (1955).

first formant. The pattern-playback experiments showed that the frequency transition in the second formant for the voiced plosive consonants (/b/, /d/, and /g/) were similar to those of Fig. 6.7(a) of the voiceless plosive consonants (/p/, /t/, and /k/), respectively. In particular, the consonant /d/ transition frequency seems to originate from 1,800 Hz as shown in Fig. 6.7(b), resembling the consonant /t/ without the first formant transition. The consonants /b/ and /g/ have similar frequency origin as that of /p/ and /k/ shown in Fig. 6.7(a).

These pattern-playback experiments also identified the formant transitions vs the place of consonant production for voiced, unvoiced and nasal consonants. In reality, the recognition of consonants /d/ /b/ and /g/ depends on the trailing vowels shown in Fig. 6.8(a). One needs a silent interval for the second formant transition to produce a clear /d/, as shown in Fig. 6.8(b). Similarly, the researchers found that the key to distinguishing the voiced /d/ from the unvoiced /t/ lies in the cue provided by the "first-formant" transition [compare Fig. 6.8(b) with Fig. 6.8(c)].

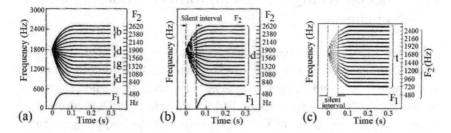

Figure 6.8: (a) The pattern extending to the /d/ focus does not always produce a clear /d/ sound; a clear /d/ sound in every case needs to erase the first part of the transition. (b) A silent plosive consonant transition originated at about 1,800 Hz and a non-transitional first formant will produce a clear /t/.

In 1976, McGurk and MacDonald discovered that visual cue also plays a role in consonant recognition, called McGurk effect, of [ga], [da], and [ba], etc.[3]

---

[3]See H. McGurk and J. MacDonald, Hearing lips and seeing voices, *Nature*, **264**, 746 (1976); doi:10.1038/264746a0

The fricative consonants are produced by forcing air through narrow gap between lips, teeth, tongue and palate. The fricative consonants /s/ and /ʃ/ ("sh") are distinguished by the "sh" sound having sound energy concentrated in the 2,000–3,000 Hz range, while /s/ is a burst of noise sound energy above 4,000 Hz, up to 8,000 Hz. The duration of a sound can also be important in phoneme recognition. A fricative consonant appears to change into a plosive when its duration is shortened, e.g. the /s/ in "see" is shortened from 100 ms to 10 ms, the word is heard as "tee."

Studies on the consonant recognition are relevant in all languages in the world. Many relevant publications are available in the literature (see Ref. [20] for comprehensive review). Recent advances in computer speech synthesis can generate sound based on spectra. Linguists are exploring methodologies for producing vocal sounds for all languages based on computer syntheses. Advances in this topic are important in language learning.

II.   Speech Analysis, Intelligibility, Modeling and Synthesis

Advances in computing technology provide a powerful tool in the fast Fourier analysis of speech. A "spectrogram" is a display of intensity vs frequency vs time in speech analysis. Figure 6.9(a) shows a three-dimensional (3D) spectrogram, called a "waterfall plot." Figure 6.9(b) shows the 2D projection of the 3D spectrogram in frequency vs time.

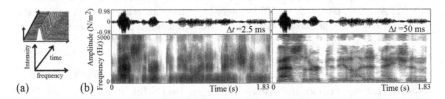

Figure 6.9: (a) A 3D spectrogram shows intensity vs frequency vs time, called the waterfall plot. (b) Waveform (top) and 2D spectrogram (bottom) vs time of W. Cronkite's newscast "ambassador to United Nations." The time window used to analyze the frequency is $\Delta t = 2.5$ ms (left plot) and 50 ms (right plot). The darkness in print (or color-coding) in the frequency spectrum represents the sound intensity.

The sound intensity is expressed in either color-code or darkness in print.

The pitch of human voice is typically about 100 Hz, 200 Hz, and 300 Hz for men, women and children. The sampling rate of voice recordings on compact disc (CD) is 44.1 kHz, i.e. the time resolution of the voice is about 23 μs. The frequency resolution in spectrogram depends on the time window Δt in the FFT analysis. If we choose $\Delta t = 2.5$ ms and 50 ms, the frequency resolutions of the FFT analysis are 32 Hz and 1.6 Hz, respectively. When the frequency resolution is small enough, horizontal strips will appear in the spectrogram. The spacing of these horizontal lines in the spectrogram is the pitch. The example shown in the right plot of Fig. 6.9(b) shows clearly the pitch of the speech. On the other hand, if Δt is much smaller than the period of the pitch, vertical strips appear on the spectrogram.

★ Example 6.3: Based on the uncertainty principle, what is the frequency resolution for the FFT sampling width of 2.5 ms and 50 ms?

Answer: According to the uncertainty principle, the frequency resolutions are $\Delta f = \frac{1}{4\pi \Delta t} = 32$ Hz and 1.6 Hz for $\Delta t = 2.5$ ms and 50 ms sampling time widths, respectively [see Eq. (4.1)]. Although 32 Hz is still smaller than the pitch of about 115 Hz, the spectrum has difficulty in identifying the pitch [see Fig. 6.9(b) left plot]. However, 2.5 ms is smaller than the 8.7 ms period of the vocal fold vibration, so the spectrograph shows the vertical lines at 8.7 ms interval. For the 50 ms sampling width, the 1.6 Hz resolution is small enough for the spectrograph to identify the horizontal lines in frequency space at the fundamental pitch of 115 Hz [see Fig. 6.9(b) right plot].

The peaks in the spectrum of a vowel sound are "formants." Figure 6.10(a) shows the spectrum of the vowel "b/a/" in the word "ambassador" of Fig. 6.9(b). Each formant has its own resonance bandwidth $\Delta f = f_2 - f_1$, where f_1 and f_2 are frequencies at 3 dB below the maximum, schematically shown in Fig. 6.10(b) [see Fig. 4.5]. The plot shows intensity vs frequency of a resonance. The resonance bandwidth is $\Delta f = f_0/Q$. Some researchers use the frequency width of 10 dB below maximum intensity for the resonance bandwidth: $\Delta f_{10} = 3f_0/Q$. Figure 6.10(c) shows the measured 10-dB formant bandwidth Δf_{10},

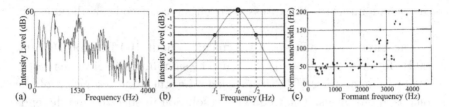

Figure 6.10: (a) The spectrum of the vowel /ba/ in ambassador vs frequency, clearly showing the formants. Each formant has a resonance bandwidth. (b) Redrawing of Fig. 4.5 in dB scale. The resonance frequency is f_0 with bandwidth $\Delta f = f_2 - f_1 = f_0/Q$, where Q is the resonance quality Q-factor. (c) The data of "10-dB Formant bandwidth" $\Delta f_{10} = 3\Delta f$ vs formant frequency compiled by G. Fant et al.

compiled by G. Fant.[4] It is about 50 Hz for the first 2–3 formants in our speech. Because the bandwidth is small, the spectrogram of a vowel shows many formants. The auditory nerve of mammals have similar frequency bandwidth in neurophysiological experiments.[5]

The first two or three formants are essential for the recognition of a vowel sound. We can still understand what is being said even when we replay a recorded speech at a higher speed such that pitch and formants are both at higher frequencies. Similarly, if a person inhales helium, formant frequencies increase without changing pitch frequency; the speech is still understandable. The speech intelligibility is an important topic in linguistic research.

A. Speech Intelligibility

Speech intelligibility depends on the speech sound level, background noise level, reverberation of the room, the spectrum of noise, and the distortion of the transmission medium. It is a complex science in the measurement of speech intelligibility. The Speech Transmission Index (STI), developed by Tammo Houtgast and Herman Steeneken in 1971, is a measure of speech transmission quality.[6] STI has gained interna-

---

[4]See H.K. Dunn, Methods of measuring vowel formant bandwidths, *Journal of the Acoustical Society of America*, **33**, 1737 (1961). Figure 9.8(c) was compiled by G. Fant, *Formant bandwidth data*, KTH-1962-3-1-001-002 (1962).

[5]See E.F. Evans, *Cochlear Nerve and Cochlear Nucleus*, DOI: 10.1007/978-3-642-65995-9_1, Springer (1975).

[6]T. Houtgast, and H.J.M. Steeneken, *Acustica*, **25**, 355 (1971).

tional acceptance as the quantifier of speech intelligibility. STI is a numeric representation measure of communication channel characteristics whose value varies from 0 to 1. Table 6.1 shows the relation of STI with intelligibility of syllables, words and sentences. Similarly, the American National Standard also provides a set of procedures for calculating the articulation index and the Speech Intelligibility Index (SII).[7] Both the STI and SII serve as the standard packages for the evaluation of voice transmission technologies such as the public address system, phone lines, voice transmission in noisy environment, noise cancellation systems, induction loop system, etc.

Table 6.1: STI value vs the intelligibility of syllabus, words and sentences.

STI value	Quality according to IEC 60268-16	Intelligibility of syllables in %	Intelligibility of words in %	Intelligibility of sentences in %
0–0.3	bad	0–34	0–67	0–89
0.3–0.45	poor	34–48	67–78	89–92
0.45–0.6	fair	48–67	78–87	92–95
0.6–0.75	good	67–90	87–94	95–96
0.75–1	excellent	90–96	94–96	96–100

A tone of low frequency more efficiently masks a tone of high frequency. A narrow-band noise is effective in masking speech if its frequency is below the speech band. In a quiet environment, typical conversation at about 30–60 dB is easily intelligible. The thresholds of intelligibility and detectability of speech vs the white or broadband noise sound level is about 10 dB above the noise level. Reduction of noise level is an important task in public environments. Technologies, such as the induction loop system, etc., can provide transmission of clear audio sounds to listeners through the magnetic induction area to their hearing aids or portable receivers in noisy environments.

---

[7]See http://sii.to/index.html for the ASA SII working group on ANSI S3.5-1997 procedures.

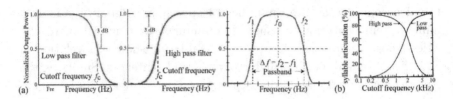

Figure 6.11: (a) Schematic drawing and definition of low-pass, high-pass, and band-pass filters. (b) The syllable articulation score vs the cutoff frequency for the high and low-pass filters.

Speech is rich in the frequency spectrum. The frequency spectrum can be filtered by a low-pass filter, a high-pass filter, or a band-pass filter, shown in Fig. 6.11(a). For the low/high-pass filters, the cutoff frequency f_{cutoff} is the frequency that the output power is $\frac{1}{2}$ of the maximum output power, or -3 dB from the power gain curve. For the band-pass filter, the center frequency and bandwidth are $f_0 = \sqrt{f_1 f_2}$ and $\Delta f = f_2 - f_1$, respectively, where f_1 and f_2 are frequencies that transmit $\frac{1}{2}$ of the maximum output power, or -3 dB in the gain curve [see Appendix B(4)].

High/low-pass filters can affect the SII of a speech sound. Figure 6.11(b) shows the articulation scores vs cutoff frequency f_{cutoff} for high- and low-pass filters. Elimination of all frequencies *below* 500 Hz in high-pass filtering will reduce the articulation score by only 2%, while the energy eliminated in this frequency range is only about 40% of speech energy. Cutoff frequency content *above* 1,500 Hz in low-pass filtering gives only about a 40% articulation score, yet the filter removes only about 10% of speech energy. The articulation score curves for high- and low-pass filtering cross at about 1,800 Hz, where the articulation score is 67% at STI value of 0.6.

For a band-pass filter, it is possible to achieve an acceptable level of articulation at a surprisingly narrow passband. The passband at $f_0 = 1,500$ Hz with $\Delta f = 1,000$ Hz has a sentence articulation score of 90%. Nevertheless, some reports indicated that a $\frac{1}{3}$-octave passband at $f_0 = 1,500$ Hz gave a sentence articulation score of 90%. A band-pass filter centered at $f_0 = 2,000$ Hz with a $\frac{1}{3}$-octave bandwidth has the intelligibility score about 50%.

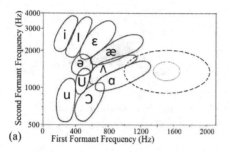
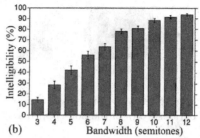

(a) First Formant Frequency (Hz)

(b) Bandwidth (semitones)

Figure 6.12: (a) Overlay in frequency space of the vowel formant map with a band-pass filter centered at 1,500 Hz at bandwidths at $\frac{1}{3}$ octave (thin solid circle) and 1,000 Hz (dashed circle) respectively. (b) Intelligibility vs the bandwidth of rectangular sharp cutoff band-pass filter obtained by Warren, Bashford and Lenz. Four semitones and twelve semitones correspond to the thin solid and dashed circles in Fig. 6.12(a) respectively.

Figure 6.12(a) shows the overlay in frequency space for the first two formants of vowels and the band-pass filter centered at 1,500 Hz with a $\frac{1}{3}$-octave bandwidth (thin solid circle) and a 1,000 Hz bandwidth (dashed circle) in the 3-dB region, respectively. Since the first-order filter only reduces 6 dB/octave, the 1,000-Hz bandwidth filter can cover the frequency space for the first two formants within the 6-dB region. Warren, Bashford and Lenz carried out experiments with various band-pass filters on intelligibility, centered at 1,500 Hz with a 100 dB/octave filter.[8] Figure 6.12(b) shows intelligibility vs bandwidth of a rectangular band-pass filter with varying bandwidth from 3 semitones to 12 semitones. At 12-semitones bandwidth, the filter has a sharp cutoff at 1,060 Hz to 2,120 Hz, and there is more than 90% intelligibility. The filter vowel covers the frequency range of the second formant but nearly excludes all first formants. It appears that our hearing system can fill in some "missing first formants," just like virtual pitch.

Modern electronic devices such as synthesizers and amplifiers may produce waveform distortion, which is a special type of filter. Peak clipping is a common distortion in electronic production of sound due to gain saturation. Figure 6.13 shows a digitized partially saturated

---

[8]R.M. Warren, J.A. Bashford Jr., and P.W. Lenz, *JASA*, **108**, 1264 (2000); see also Ref. [20].

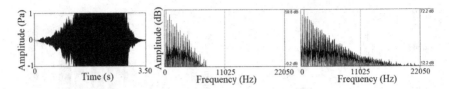

Figure 6.13: Waveform distortion by peak clipping in a violin E4 note and its Fourier spectra of the non-clipped and highly clipped sections.

waveform of a violin E4 sound in crescendo and the FFT spectra of the unsaturated and saturated sections. Note that the pitch remains the same, while the peak clipping part produces many higher harmonics and changes the formants. However, clipping plays a lesser role in impeding sound intelligibility.

★ Example 6.4: Find the lower and upper cutoff frequencies of a pass-band centered at 1,500-Hz with 1,000-Hz bandwidth, and the lower and upper cutoff frequencies of the octave band.

Answer: Using Eq. (A.9) with $f_c = 1,500$ Hz and $\Delta f = 1,000$ Hz, we find $f_1 = 1,081$ Hz and $f_2 = 2,081$ Hz. For the octave band centered at 1,500 Hz, the lower and upper frequencies are respectively at $f_1 = 1,500/2^{1/2} = 1,060$ Hz and $f_2 = 1,500 \times 2^{1/2} = 2,120$ Hz. Note that these two bands cover nearly identical frequency range. The octave band is the band of 12 semitones.

B.   Modeling and Synthesis of Speech

Human speech is a curiosity in human evolution. Many scientists and engineers have tried to replicate the intricate human speech system. In 1779, Christian Kratzenstein built models of the human vocal tract that could produce vowel sounds. In 1791, Wolfgang von Kempelen of Pressburg, Hungary built a talking machine with tongue and lips. In 1837, Charles Wheatstone reconstructed von Kempelen's talking machine with a leather resonator that could be manually modified for phoneme production. Alexander Graham Bell also built a von Kempelen talking machine. In 2010, engineers at Kagawa University in Japan used an air pump, artificial vocal cords, a tunable resonance tube, and a nasal cavity to represent human vocal organs.[9] Additional microphones can analyze formants of vowel sounds for feedback.

Modern talking machines use electronic sound generators rather than mechanical models. Two important characteristics of speech synthesis systems are naturalness and intelligibility. A "speech synthesizer" is a computer system used for human speech production. The British film "2001: A Space Odyssey" introduced the voice of artificial intelligence computer system HAL-9000. This AI-voice movie score was motivated by J.L. Kelly's speech-synthesis demonstration system using the IBM-704 computer at Bell Labs. Kelly's voice recorder synthesizer, or *vocoder*, recreated the folksong "Daisy Bell (Bicycle built for two)."

Since the 1980s, there are text-to-speech computer programs in almost all languages. Phone call centers are equipped with voice recognition computers and are able to provide intelligent interaction with callers. Many intelligent robots are available for commercial use. They can communicate with humans in different languages and provide valuable services to human society. In the late 20th century, computing power increased exponentially. In 1997, "Deep Blue" won the chess championship over Garry Kasparov. In 2011, IBM's learning, human-aware computer "Watson" won a competition in "Jeopardy" against quiz masters Ken Jennings and Brad Rutter. The artificial intelligence (AI) system could comprehend human language and provide answers to questions. Scientists and engineers are developing AI systems that can provide services in health and medicine, social interaction, and scientific research.

III. Singing

Human beings are born with an incredible music box in our bodies. For singing, our vocal organ can produce accurate pitch and change that pitch in an instant. The pitch of an adult male voice is usually lower due to larger vocal folds, while the pitch of female voice is higher due to smaller vocal folds. We classify singing voices into bass, baritone,

---

[9]See Toshio Higashimoto and Hideyuki Sawada, *Proceedings of 2002 IEEE Conference on Robotics and Automation*, 3858, Washington DC; DOI: 10.1109/ROBOT. 2002.1014322

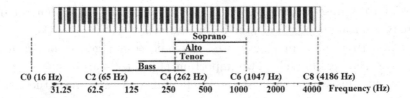

Figure 6.14: Voice range vs the 88-key piano frequency range.

tenor and countertenor (ranging from E2 to even F6) for men, and contralto, mezzo-soprano and soprano (ranging from F3 to C6) for women. General vocal ranges are soprano in C4–C6; mezzo-soprano in A3–A5; contralto in F3–F5; tenor in C3–C5; baritone in F2–F4; and bass in E2–E4. The fundamentals of human voices are roughly in the range of 80 Hz to 1,100 Hz (that is, F2 to C6) for normal male and female voices. Figure 6.14 lists some typical singing frequency ranges.

A.   Singing Pitch and Formants

The vocal folds control pitch, and the vocal tract determines the formants that articulate the sound with vowels and consonants. Singing vowels are similar to the spoken vowels. Figure 6.15 shows a schematic drawing of vowel sound /ah/ at 100 Hz and 220 Hz. Since the first formant of /ah/ is around 650 Hz, if the fundamental pitch were above 650 Hz, the singing voice would miss the first formant of the vowel sound /ah/. Figure 6.5(b) overlay the singing range (dashed-line box)

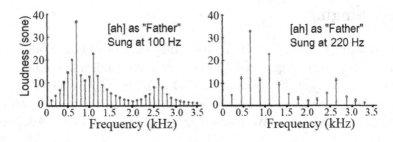

Figure 6.15: Schematic drawing of spectra loudness vs frequency for the vowel /ah/ sung at 100 Hz and 220 Hz, assuming the same formant in singing and in speech (see Ref. [31]).

with the first two formants of vowel sounds (see Ref. [31]). When the singing pitch is higher than the first formant of a vowel, the singing note will miss the first formant of the vowel.

B.   Formant Tuning in Singing and Singer's Formant

In normal speech, we do not artificially tune our formant to match a harmonic of the fundamental frequency for vowel sounds. However, expert professional singers can tune their formant frequencies to produce a louder and brighter singing note. Experienced singers have the skill to tune one of their formants to coincide with the frequency of the sung note. Figure 6.16(a) shows the formants F1 (solid line) and F2 (dashed line) of vowels /u/, /o/, /a/, /e/, /i/ and /y/ sung by a professional singer vs the pitch frequency and its partials (dotted line), as measured by J. Sundberg (see Ref. [31]). One of the formants of all the vowels converges to one of the harmonics of the pitch. It seems that all vowels, except /o/, have the same formant at high pitch frequency in singing. It may sometimes be difficult to distinguish vowels because of the formant tuning. It seems that musical presentation is more important than the actual text in opera singing. Sundberg also discovered the singer's spectral sound level enhancement at 2,500 to 3,000 Hz, shown

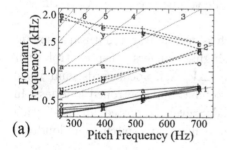 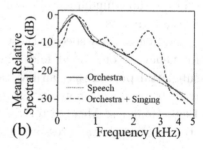

Figure 6.16: (a) The first formant (solid line) and second formant (dashed line) of vowels /u/, /o/, /a/, /e/, /i/ and /y/ vs the sung pitch of a soprano, as measured by Sundberg. The dotted lines are partials of the sung pitch, marked in numbers. At 700 Hz, formants of all singing vowels converge to harmonics of the pitch. (b) The singer's formant stands out over the spectral intensity level of orchestra and spoken voice at 2,500–3,000 Hz (see Ref. [31]).

in Fig. 6.16(b).[10] He conceived the "singer's format" as an "additional vocal resonance" in the vocal box by lowering the larynx. He modeled the vocal box as a closed tube of about $\frac{1}{6}$ of the length of the vocal track that produces resonant enhancement at frequency in the 2,500 to 3,000 Hz range.

★ Example 6.5: The singer's formant in Fig. 6.17(b) is at 2,700 Hz. What length of the larynx lowering needed to achieve this resonance frequency? (Assuming the speed of sound is 350 m/s at the larynx.)

Answer: The resonances for the closed cylindrical pipe is $f_1 = \frac{v}{4L} = \frac{350}{4 \times L} = 2,700$ Hz. The length of the larynx box is $L = 3.2$ cm.

Analyses by D.G. Miller found that Pavarotti chose the second resonance formant tuning while Domingo chose the singer's formant to enhance an identical B4^b note in opera singing (see Ref. [27]). He pointed out that tools such as the electroglottalgraph (EGG) and the acoustic waveform analysis are useful in training singers on formant tuning. An extreme formant tuning is the throat singing of Tuva, Mongolia, Tibetan and other African tribes.[11] The singers can collapse several formants into a single frequency that may also coincides with the harmonics of the fundamental frequency for achieving nearly pure sinusoidal-like tone in ceremonial chants. Expert throat singer can maintain the same fundamental pitch while tuning the frequency of the collapsed formants.

During normal speech, the pitch is normally the natural resonant frequency of our vocal folds. In singing, singers must control their subglottal pressure p_s, airflow rate ($\Delta V/\Delta t$), vocal cords, and vocal tract to produce a note at varying pitches and controlled loudness. In 1986, Bouhuys, Mead, Proctor, and Stevens reported an interesting study of subglottal pressure, volume flow rate, pitch, and loudness

---

[10] J. Sundberg, Formant technique in a professional female singer, *Acta Acustica*, **32**, 89 (1975); J. Sundberg, The acoustics of the singing voice, *Sci. Am.*, **236** (1977); see also Ref. [31].

[11] T.C. Levin and M.E. Edgerton, *The Throat Singers of Tuva* (*Scientific American*, 1999); see also https://www.youtube.com/watch?time_continue=60&v=Mtbsb3swbcU.

level in singing.[12] They measured the loudness level with a microphone three inches in front of the singer's mouth for trained and non-
trained people. They discovered that good quality singing obeyed the
condition $p_s \sim (\Delta V/\Delta t)$, independent of sound level, while the sound
intensity level proportional to p_s^3. Since the power needed in singing is
$p_s \times (\Delta V/\Delta t) \sim p_s^2$, the efficiency is higher at louder singing. Bouhuys'
measurements showed that it took about 3,500 Pa with volume flow
rate of 400 cm^3/s to produce 116 dB of sound intensity. The efficiency of sound production is typically 0.1%–2% (see Example A.7 in
Appendix C).

IV. Summary

The human vocal organ is similar to a musical instrument composed of
a sound source and a resonant chamber. The speech sound source arises
from airflow driven by the lungs. The periodic air puffs, generated by
the vibrational vocal cords in the larynx, produce a buzzing sound rich
in frequency content. The resonant chamber is composed of the vocal
tract, which includes the larynx, pharynx, nasal cavity, and oral cavity.
It modulates the buzzing sound of the glottal airflow into vowels. We
differentiate each vowel by its particular formants. The sound produced
by rapidly changing airflow with air constriction and sudden release at
various locations along the vocal tract, e.g. velum, alveolus, teeth and
lips, generates consonants. The vowels and consonants of speech are
universal in all languages. Each is called a "phoneme," the basic unit of
articulated speech sound. The agility of both the tongue and vocal folds
can provide wide tuning of the voice and frequency in a very short time.
Beyond speech sounds, lip reading, facial expression and hand gestures
are also integral parts of human communication.

  The production of singing sound is the same as that of speech.
Experienced professional singers may employ formant tuning or singer's
formant to enhance the beauty and loudness of a singing note. Formant tuning is applicable to either chest (regular) voice singing or head

---

[12]Arend Bouhuys, Jere Mead, Donald F. Proctor, and Kenneth N. Stevens,
Annals of the New York Academy of Science, 1968; DOI: 10.1111/j.1749-6632.
1968.tb56760.x

(lighter) voice singing, known as falsetto singing. The extreme of formant tuning is the throat singing of Tuva, where all formants may collapse into one frequency for a nearly pure sinusoidal tone.

Professional singers, actors, sport casters, and teachers put a lot of stress on their vocal folds. Stressful overuse of the vocal folds can lead to scar tissue on vocal folds. Surgery and innovative technology can restore a professional singer's career, as reported by John Colapinto in *The New Yorker* magazine: "A surgeon pioneers methods to help singers sing again."[13]

Advance in vocal data acquisition and analysis such as the glottal stroboscope, X-ray imaging, magnetic resonance imaging (MRI), electroglottalgraph (EGG), electromyography (EMG), and acoustic waveform synthesizer can provide research and diagnosis in speech, language learning and singing performance improvement. Advances in computer technology provide artificial intelligent (AI) systems the ability to synthesize and understand the syntax of human language. An AI system may someday be able to sing and imitate a particular person's human voice. It is scary.

V.   Homework 06

1. The human voice and the human ear match very well. The ear has its maximum sensitivity in the frequency range from _____ Hz to _____ Hz and that are the same resonant frequency range of the vocal tract.

2. Speech sound is a product of the source, _____ flow; the filter, _____ tract; and the radiator, _____ opening.

3. Speech sounds originate in the _____ that modulates the airflow with vibrations of the _____ _____ .

---

[13] John Colapinto, *Giving Voice*, https://www.newyorker.com/magazine/2013/03/04/giving-voice

4. Resonances of the vocal tract, called _____ , are tuned by changing the length, cross-sectional area and volume of the vocal tract.

5. A display of intensity vs frequency in speech analysis is a

_____ .

6. The voice spectra for two different sounds, shown below, have a fundamental pitch of 125 Hz.

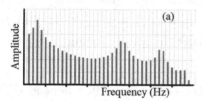 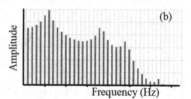

The formant frequencies for spectrum (a) are: $f_1 =$ _____ Hz, $f_2 =$ _____ Hz, and $f_3 =$ _____ Hz. The formant frequencies for spectrum (b) are: $f_1 =$ _____ Hz, $f_2 =$ _____ Hz, and $f_3 =$ _____ Hz.

7. [Choose one (a) or (b)]. The voice spectra (a) and (b) below are IPA vowel /i/. Each vertical line under the envelope curve represents the strength of a harmonic of the fundamental pitch. Which spectrum represents a female speech?

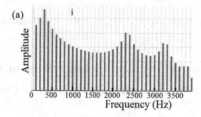 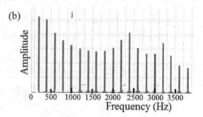

8. A complex waveform pattern repeats itself every 5.0 ms. The frequency of the fundamental is _____ Hz and the next three Fourier components are _____ Hz, _____ Hz, and _____ Hz.

9. The individual units of sound that make up speech are
   _____ . They are divided into two groups: _____ that
   are voiced and _____ that are voiced or unvoiced.

10. The first two or three _____ are usually sufficient for the
    recognition of vowel sounds.

11. The simplest acoustic model of the vocal tract is a pipe closed at
    one end (by the glottis) and open at the other end (at the lips). For
    a pipe with a length $L = 17$ cm, the first three allowed resonances
    are: $f_1 =$ _____ Hz, $f_3 =$ _____ Hz, and $f_5 =$
    _____ Hz. (Assume that the speed of sound in the mouth
    to be about 340 m/s.)

12. The cues for the recognition of consonants often depend on the
    _____ sound that follows.

13. A speech spectrogram is normally a 3D plot of sound level versus
    _____ , and _____ .

14. A (subglottal) pressure of 4000 N/m^2 of a singer singing an A3 note
    at 220 Hz can produce a sound intensity level of about 120 dB at the
    mouth opening. The sound intensity, I (in W/m^2), that corresponds
    to this sound loudness level (in dB) = _____ W/m^2.
    Assume that the area of the mouth opening is 20 cm^2 or 2.0×10^{-3}
    m^2. The total radiated sound power is _____ W. If
    each air-pulse has a volume of 2 cm^3, the power in singing is
    _____ W.

15. Sung vowels and their formants differ from spoken vowels in the
    appearance of a "singer's formant" in the frequency range between
    _____ Hz and _____ Hz.

16. The power (in watts) used to move air out of the lungs is equal
    to the air pressure (in N/m^2) multiplied by the airflow rate
    (in m^3/s). For quiet breathing (where $p = 100$ N/m^2 and flow

rate $= 100$ cm^3/s or 1.0×10^{-4} m^3/s), the power involved $=$
_____ W. For loud singing (where $p = 4{,}000$ N/m^2 and
flow rate $= 400$ cm^3/s or 4.0×10^{-4} m^3/s), the power involved is
_____ W.

17. The intensity of fortissimo singing at 120 dB measured at 3 inches
    from the singer's mouth. What is the power of the sound source
    from the mouth? (Assuming uniform spreading from the singer's
    mouth.)

18. For loud singing, the subglottal pressure $p_s = 4{,}000$ N/m^2 and
    flow rate at 400 cm^3/s will produce an SIL of 120 dB measured
    at 3 inches from the mouse opening. What is the efficiency in loud
    singing? (See Eq. (A.4) for power needed to move air.)

19. A trained singer can release 4 liters of air from his/her lungs after
    each deep breath. From the triangular waveform, we can estimate
    that each puff of air contains 2.0 cm^3 (2.0×10^{-3} liters). If the
    singer sounds the E3 note (165 Hz), (a) the number of air puffs is
    _____ emitted per second; (b) the singer can sustain the
    note for _____ seconds on a single breath at this rate.

    It is tempting to think that a singer sings E4 note at 330 Hz would
    double the volume flow rate. Measurements show that the volume
    flow rate of "trained singers" remain nearly constant, i.e. when
    increasing the frequency, the singer will accordingly reduce the
    volume of each air puff. The correlation between subglottal pressure
    and the pitch in singing is still ambiguous (see Ref. [30]).

20. Assume that a "singer's formant" at 3,000 Hz is due to a resonance
    on part of the larynx above the vocal cords. Using a cylindrical-
    closed-pipe model for the upper-larynx tube, we estimate the
    length of this region to be _____ m or _____ cm.
    (Assuming speed of sound in the mouth to be 350 m/s.)

21. Based on Sundberg's proposal, we model the larynx as a short
    closed cylindrical pipe of length of 3 cm for the singer's formant
    [see Fig. 6.1(a)]. The resonance frequency is _____ Hz
    (Assume the speed of sound wave in the vocal box be 350 m/s.)

Chapter 7

Electricity and Magnetism in Acoustics

Electricity and magnetism (E&M) are important in acoustics as well as in daily life. We apply E&M technologies to sound production, manipulation, and communication. This chapter discusses some basic properties of electricity and magnetism, electric circuits, electronic components, amplifiers, filters, generators, oscillators, transducers, transformers, loudspeaker, and microphones. All of these topics are relevant to music and sound.

I. Electric Force, Electric Current, Electric Potential, and DC Circuit

A. Coulomb's Law

Matter is made of atoms. Each atom is made of electrons orbiting around its atomic nucleus. A nucleus is composed of protons and neutrons bound together by a strong nuclear force. Electrons and protons carry "charge," while neutrons carry no charge. An electron carries a negative charge of $-\mathbf{e}$, and a proton carries $+\mathbf{e}$, where the fundamental charge unit is $\mathbf{e} = 1.602 \times 10^{-19}$ [coulomb] or [C]. Charge particles interact via Coulomb's law. The magnitude of Coulomb's force is

$$F = k\frac{q_1 q_2}{r^2}, \qquad (7.1)$$

where $k = 8.99 \times 10^9 \text{ N} \cdot \text{m}^2/\text{C}^2$, and r is the distance between the two point charges q_1 and q_2. Charges of the *same sign repel* each other, while those of *opposite sign attract* each other. A nucleus is composed of Z number of protons and N number of neutrons. A neutral atom will also Z number of electrons outside the nucleus, and thus a neutral atom carries no charge. The number of protons Z is the "atomic number," which determines chemical properties of the atom.

★ Example 7.1: The mass of proton and electron are 1.67×10^{-27} kg and 9.11×10^{-31} kg, respectively. The average distance between the electron and the proton in the hydrogen atom is $r = 5.29 \times 10^{-11}$ m. Compare the electric force vs gravitational force of the hydrogen atom.

Answer: From Eq. (1.4), the gravitational force between the proton and electron is

$$F_{grav} = (6.67 \times 10^{-11} \ \text{Nm}^2/\text{kg}^2)(1.67 \times 10^{-27} \ \text{kg})(9.11 \times 10^{-31} \ \text{kg})/$$
$$(5.29 \times 10^{-11} \ \text{m})^2 = 3.63 \times 10^{-47} \ \text{N}.$$

On the other hand, the Coulomb force between the electron and proton in a hydrogen atom is

$$F_{Coul} = (8.99 \times 10^9 \ \text{Nm}^2/\text{C}^2)(1.602 \times 10^{-9} \ \text{C})^2/(5.29 \times 10^{-11} \ \text{m})^2$$
$$= 8.24 \times 10^{-8} \ \text{N}.$$

The Coulomb force is much stronger than the gravitational force, with $F_{coul}/F_{grav} = 2.27 \times 10^{39}$. In other words, Coulomb force is 10^{39} times greater than the gravitational force. The proton and the electron in the hydrogen atom are bound together mainly by the Coulomb force.

B.   Electric Potential; Electromotive Force (EMF) and Battery

Electric potential energy is the energy associated with a charge's position due to Coulomb's force. The electric potential, denoted by "V," is the electric potential energy per unit charge at a position. Thus, the electric potential energy U is equal to $U = q \times V$, where q is the charge and V is the electric potential. The unit of the electric potential is [Volt] or [V]. The product of [C] and [V] is [joule] or [J], an energy unit.

Positive charges will move from a high electric potential location to a place at low electric potential, just like "an object will fall from a higher position to a lower position due to the gravitational force on Earth." Batteries and AC generators are an *electromagnetic motive force* (EMF) or power source that maintains electric potential between two terminals. A battery maintains a constant electric potential by converting chemical energy to a constant electric potential until the chemical energy depleted. The symbol for battery is ⊣⊢, with the long side designated as the *positive* potential terminal. The potential in a battery is in the unit of [V]. Typical common commercial batteries have $1.5 \sim 3$ [V] per cell. One can connect batteries in series to achieve a higher potential [voltage] or connect them in parallel to achieve a higher capacity for longer running time.

C.   Electric Current

A positive charge can move from a high-potential position to a low-potential position. When charge moves, the rate of total charge per unit time is the electric current, i.e.

$$ I \equiv \frac{\Delta Q}{\Delta t} . \tag{7.2} $$

The SI unit of electric current is [ampere] or [A], where 1 [A] = 1 [C/s]. The charge carrier in an electric circuit is normally electrons. They move in the opposite direction.

D.   Ohm's Law: $V = I \cdot R$

As electric charges move in medium, electrons (charge carriers) may encounter "resistance" from atomic lattices in electronic elements. Conductors are metals in which electrons can move with "little" or no resistance. The symbol for the conductor is a solid-line ——. Resistors are electronic elements that partially conduct electricity. They are made of specially designed materials in which electric current may encounter varying degree of resistance. The symbol of resistor is -\/\/\/-. Ohm's law governs the motion of charges in resistors:

$$ V = I \cdot R , \tag{7.3} $$

where V is the "voltage drop" across the resistor, I is the electric current, and R is the resistance. The units of V, I and R are [volt], [ampere] and [ohm or Ω], respectively.

Electronic elements that obey Ohm's law are *"Linear,"* i.e. the electric current is proportional to the electric potential drop. Resistors, capacitors, and inductors are linear, because they obey Ohm's law. We can use resistors as current limiters, voltage divider, and potentiometer in a circuit. Resistors can function as heat sources, such as incandescent light bulb or hot plates for heating and cooking. In an electric circuit, we can connect resistors in series or in parallel.

E.   Resistors Connected in Series or in Parallel

We can connect resistors in series or in parallel. Figure 7.1(A) shows the equivalent circuit of resistors in series and in parallel. Ohm's law of the equivalent circuit is $V_{ab} = IR_{eq}$, where V_{ab} is electric potential between terminals a and b, R_{eq} is the equivalent resistance, and I is the electric current passing through the equivalent resistor.

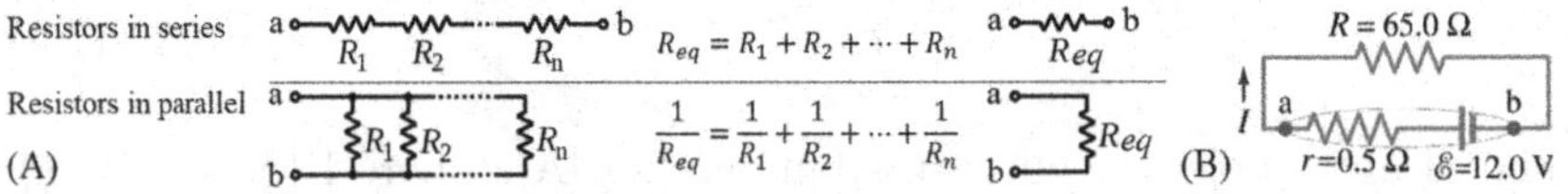

Figure 7.1: (A) Resistors connected in series (top) or in parallel (bottom) to terminals a and b. (B) A simple circuit with a 65 Ω resistor connected to the battery terminals a and b. The battery has its own *internal* resistance at 0.5 Ω, in series with the 65 Ω resistor.

When we connect resistors in series, the electric current will pass through all resistors. The total voltage drop is the sum of the voltage drops in each individual resistors, and the equivalent resistance R_{eq} of the equivalent circuit is the sum of resistance of all resistors. When resistors are connected in parallel, a different amount of current passes through each resistor to maintain the same terminal voltage, i.e. $V_{ab} = I_1 R_1 = I_2 R_2 = \cdots = IR_{eq}$. Since the total current is $I = I_1 + I_2 + \cdots$, the inverse of the equivalent resistance is the sum of the inverse resistance of all resistors:

$$\text{Resistors in series:} \quad R_{eq} = R_1 + R_2 + R_3 + \cdots \quad (7.4a)$$

$$\text{Resistors in parallel:} \quad \frac{1}{R_{eq}} = \frac{1}{R_1} + \frac{1}{R_2} + \frac{1}{R_3} + \cdots \quad (7.4b)$$

Figure 7.1(B) shows an example of a circuit, where the same electric current passes through an internal resistor r of the battery and the external resistor R in series. The equivalent resistance is $R_{eq} = 65.0 + 0.5 = 65.5\ \Omega$, and the current is $I = 12.0/65.5 = 0.183$ A.

★ Example 7.2: In circuit (A) below, the battery EMF is $V = 24$ volts and the resistors are $R_1 = 24\ \Omega$ and $R_2 = 48\ \Omega$. What is the equivalent resistance of the circuit, and what is the current I? In the circuit (B) below, the battery EMF is $V = 24$ volts and the resistors are $R_1 = 24\ \Omega$ and $R_2 = 48\ \Omega$. What is the equivalent resistance R_{eq} of the circuit, and what is the current I, I_1 and I_2?

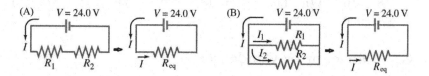

Answer:

(A) Since R_1 and R_2 are connected in series, $R_{eq} = 24 + 48 = 72\ \Omega$. The current, passing through both resistors R_1 and R_2 in the circuit, is $I = 24\,\text{V}/72\,\Omega = 0.333$ [A].

(B) Using Ohm's law, we find $I_1 = \frac{24\,\text{V}}{24\,\Omega} = 1$ [A] and $I_2 = \frac{24\,\text{V}}{48\,\Omega} = 0.5$ [A]. The total current is $I = I_1 + I_2 = 1.5$ [A]. The equivalent resistance of the equivalent, circuit shown at right plot in (B), is $V = I R_{eq}$, or $R_{eq} = 24\,\text{V}/1.5\,\text{A} = 16\ \Omega$. Using Eq. (7.4b): $\frac{1}{R_{eq}} = \frac{1}{24} + \frac{1}{48} = \frac{3}{48}$, we find $R_{eq} = 16\ \Omega$. The current is $I = 24\,\text{V}/16\,\Omega = 1.5$ [A], the same as what we have obtained above.

F.   Electric Power and Energy

By definition, the electric potential energy is *charge* times the electric *potential*. Power is the rate of change of the energy, and the current is the rate of charge flowing through. Thus, the power is *current* times the electric potential. With Ohm's law, we find that power has equivalent forms

$$P = IV = I^2 R = \frac{V^2}{R}. \qquad (7.5)$$

Naturally, by definition, we have (Energy) = (Power) × (time). For example, the electric power companies charge the kWh electric energy used, where 1 kWh = 1000 W × 3600 s = 3.6 MJ energy.

★ Example 7.3: An average laptop consumes about 25 W of power in operation (typically 5–50 W). What is the total power if there are 40,000 computers working simultaneously on a campus? What is the energy used and cost of electricity in 1 year if the energy cost is $0.08/kWh?

Answer: The total power of all computer is 25 W × 40,000 = 1,000 kW. The energy used in 1 year is 1,000 kW × (365 × 24 h) = 8.8 × 10^6 kWh. The cost is (8.8 × 10^6 kWh) × ($0.08/kWh) = $700,000.

G. Capacitor or Condenser

The capacitor (or condenser) is a device that can store electric energy. Capacitors are made of two parallel metallic plates or sheets separated by some non-conducting materials. When one of the plates is positively charged, the opposite metallic plate will carry an equal amount of negative charge. This process stores electric charges on the device, and it stores electric potential energy. This device is a capacitor or condenser. The symbol for a capacitor is ⊣⊢ . The "capacitance" is the ratio of stored charge q to the electric potential between two plates, i.e. $C \equiv q/V$. The capacitance of two parallel plates with area A separated by a distance d with insulating material is

$$C = \varepsilon A/d, \tag{7.6}$$

where ε is the dielectric constant of the insulator between these two parallel sheets. The unit of capacitance is [Farad]. Typical magnitude of capacitance is [pico-Farad (pF)], [nano-Farad (nF)], or [micro-Farad (μF)]. The capacitance of a capacitor at [μF] or above is considered large.

II. AC Circuit, AC Current, and AC Electric Potential

A battery converts chemical energy to electromotive force, which can drive charges to move in a circuit in one direction. This type of circuit is a direct current (DC) circuit. When the electromotive force (EMF) and voltage oscillate in time, the current also oscillates in the electric circuit. This is an alternating current (AC) circuit. The symbol for an AC voltage supply is –⊘–. Common electric generators produce sinusoidal EMF with $V = V_{\max} \sin 2\pi f t$, where $V_{\max}$ is the amplitude or maximum voltage, $f = 1/T$ is the frequency, and T is the period. Figure 7.2 shows a schematic drawing of an AC voltage. The peak-to-peak AC voltage is $V_{\mathrm{pp}} = 2V_{\max}$.

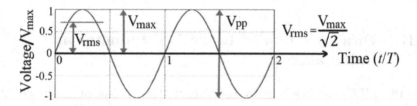

Figure 7.2: Schematic drawing of sinusoidal alternating voltage source, where $V_{\max}$ is the amplitude, V_{pp} is the peak-to-peak voltage, and V_{rms} is the root-mean-square voltage.

In US home, the AC electric voltage is $V_{\max} = 170$ [V] or $V_{\mathrm{rms}} = 120$ [V] at frequency $f = 60$ Hz. The AC power supply can also provide 3-phases with 3–4 wires offsetting each other by 120° phase to achieve a more stable machine operation in industrial applications. Outside the USA, $V_{\max}$ may vary from 170 to 330 V with frequency 50 Hz or 60 Hz.

The frequency of an AC circuit in modern electronic circuits can vary from 0 (DC) to very high frequency. The band of 20 Hz to 20,000 Hz is the audio frequency, 20 kH – MHz is ultrasound, and the band of 3 kHz to 300 GHz is radio frequency (RF), used in AM/FM radio, television, wireless communication, cell phones, satellite communication systems, etc., and the band of 100 MHz to 300 GHz is microwave. The root-mean-square (rms) AC electric potential value is $V_{\mathrm{rms}} = \frac{V_{\max}}{\sqrt{2}}$, and the average power dissipated in a circuit is

$$P_{\text{av}} = I_{\text{rms}} V_{\text{rms}} = I_{\text{rms}}^2 R = \frac{V_{\text{rms}}^2}{R} = \frac{1}{2} P_{\text{peak}} . \qquad (7.7)$$

★ Example 7.4: An amplifier under test can drive a sinusoidal signal with an rms amplitude of 4 V. When connected to an 8 Ω loudspeaker, what is the *average power* delivered by the speaker?

Answer: The average power is related to rms amplitude by $P_{\text{average}} = V_{\text{rms}}^2/R = 4^2/8 = 2$ W.

★ Example 7.5: An incandescent light bulb with a 24 Ω filament resistor is connected to an AC power line at $V_{\text{rms}} = 120$ volts. What is the *average power* consumed in the light bulb?

Answer: The average power is $P_{\text{average}} = V_{\text{rms}}^2/R = (120)^2/24 = 60$ W.

III. Faraday's Law of Induction; Inductor and Inductance

In 1820, H.C. Oerstead discovered that electric current on a wire could generate magnetic field to move the compass needle. This discovery linked electricity and magnetism into *electromagnetism*. In 1831, M. Faraday discovered the magnetic induction law: *varying magnetic field could also generate electric field*, called Faraday's law of induction. The induced EMF is

$$\varepsilon = -\Delta\Phi/\Delta t , \qquad (7.8)$$

where $\Delta\Phi/\Delta t$ is the rate of change of magnetic flux Φ, defined as the product of the magnetic flux density (loosely called magnetic field) and the area through which the magnetic field is passing. The unit magnetic flux is [tesla-meter^2] or [T-m^2]. The negative sign $(-)$ signifies that the induced EMF is "against" the change of the magnetic flux.

A. Inductor

When the electric current flows in a wire, it produces a magnetic field around the wire (Oerstead's discovery). When coiling the electric wire together, the magnetic field concentrates in the coil section. The device stores magnetic field or magnetic energy. This device is an inductor with symbol —〰〰〰—. The total magnetic flux in the inductor is proportional to the current. The inductance L is the amount of total magnetic flux increase per unit increase of current, i.e.

$$L \equiv \Delta\Phi/\Delta I. \tag{7.9}$$

The unit of inductance is the henry [H]. Typical inductance values of inductors range from millihenry [mH] to henry.

B. Transient Effect of Capacitor and Inductor in a DC Circuit

The electric current in a DC circuit with capacitors or inductors obeys Ohm's law. However, the current and voltage show "transient effects." Figure 7.3(a) shows an example of connecting a capacitor to a DC voltage source. When the switch is turned on at time $t = 0$, the current begins to flow in the circuit and charges the capacitor up to a maximum voltage across the capacitor, up to the EMF of the battery. Then the current stops. The time constant of the circuit is $\tau_C = RC$, i.e. the voltage across the capacitor is $V_C(t) = \varepsilon(1 - e^{-t/\tau_C})$, where ε is the EMF voltage of the battery (see Appendix B(6) for the exponential function). The energy stored in the capacitor is $\frac{1}{2}C\varepsilon^2$.

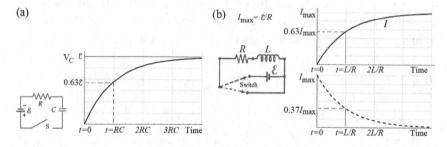

Figure 7.3: (a) A capacitor connected to a battery. When the switch S is connected, the voltage across the capacitor increases until it reaches the EMF of the battery. (b) A resistor and an inductor, connected in series to a battery. When the switch is turned on, the current is initially zero until it reaches a maximum current $I_{max} = \gamma\varepsilon/R$. If the inductor is switched to bypass the battery, the current will decay to zero.

Figure 7.3(b) shows an inductor and a resistor connected in series to a DC circuit. When the switch of the RL circuit is turned on, the current begins to flow into the inductor. The induced EMF across the inductor due to Faraday's law of Eq. (7.8) is initially maximally against the increase of the current. Thus, the current is 0 at time $t = 0$. Gradually, the induced EMF across the inductor reduces, and the current vs time t is $I(t) = \frac{\varepsilon}{R}(1 - e^{-t/\tau_L})$, where the time constant is $\tau_L = \frac{L}{R}$. The amount of magnetic energy stored in the inductor is $\frac{1}{2}LI^2$. As time $t \to \infty$, the induced EMF becomes zero because the rate of change of magnetic flux is 0, and the current reaches $I_{\max} = \varepsilon/R$, as if there is no inductor. Now, when the switch is suddenly switched to the conductor side without the battery, the magnetic energy stored in the inductor begins to dissipate in the resistor. The current, starting from its maximum current at ε/R, will decay to zero, i.e. $I(t) = \frac{\varepsilon}{R}e^{-t/\tau_L}$.

C. Ohm's Law in AC Circuits

When we connect inductors and capacitors to AC voltage source, Ohm's law also works:

$$V_{\mathrm{rms}} = I_{\mathrm{rms}}Z \text{ with } Z_C = \frac{1}{2\pi fC} \text{ and } Z_L = 2\pi fL, \qquad (7.10)$$

where Z (Z_C or Z_L) is the impedance or the reactance, measured in [ohm]. Figure 7.4 shows schematic drawings of a capacitor and an inductor in AC circuits. In the capacitor circuit, the current leads the capacitor voltage by 90°, or the voltage lags the current by 90°. In the inductor circuit, the current lags the inductor voltage.

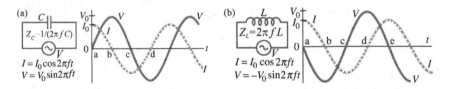

Figure 7.4: (a) Capacitor in an AC circuit. The current leads the voltage by 90° phase. (b) Inductor in an AC circuit. The current lags behind the voltage by a phase of 90°.

★ Example 7.6: An inductor with inductance $L = 5$ H is connected to an AC source with frequency $f = 60$ Hz and $V_{rms} = 120$ V, what is the reactance and the rms current of the circuit?

Answer: The impedance is $Z = 2\pi \times 60\,\text{Hz} \times 5\,\text{H} = 1885$ Ω; and the current is $I_{rms} = 120/1{,}885 = 0.064$ A.

★ Example 7.7: A capacitor with capacitance $C = 5$ μF is connected to an AC source with frequency $f = 60$ Hz and $V_{rms} = 120$ V, what is the reactance and rms current of the circuit?

Answer: $Z = 1/[2\pi \times (60\,\text{Hz}) \times (5 \times 10^{-6}\,\text{F})] = 530$ Ω; $I_{rms} = 120\,\text{V}/530\,\Omega = 0.23$ A.

★ Example 7.8: When connected to an AC generator at 60 Hz, the reactance of a capacitor is 55 Ω. What is its capacitance?

Answer: Using $Z = 55\,\Omega = 1/(2\pi \times 60\,\text{Hz} \times C)$, we find $C = 48$ μF.

★ Example 7.9: A 100×10^{-6} F capacitor is connected to an AC generator with $V_{rms} = 33.0$ V and a frequency of 200 Hz. What is the rms current in this circuit?

Answer: The impedance is $Z = 1/(2\pi f C) = 1/(2 \times 3.14 \times 200 \times 100 \times 10^{-6}) = 8.0$ Ω, and we find $I_{rms} = V_{rms}/Z = 33.0/8.0 = 4.1$ A.

★ Example 7.10: At what frequency will a generator with $V_{rms} = 120$ V produce $I_{rms} = 10$ mA in a capacitor with $C = 0.015 \times 10^{-6}$ F?

Answer: $Z = V_{rms}/I_{rms} = 12{,}000$ Ω, We find $f = 1/(2\pi Z C) = 1/(2 \times 3.14 \times 12{,}000 \times 0.015 \times 10^{-6}) = 880$ Hz.

D.   Transformer

An inductor winding can store magnetic flux and magnetic energy. Combining this mechanism with some magnetic materials that can confine and transport magnetic flux at little loss, one can design a transformer, schematically shown in Fig. 7.5. The transformer can amplify or reduce the electric potential in the secondary winding. It is an electromagnetic device used to step up/down AC voltage. The input AC voltage produces magnetic flux in the winding, and the varying magnetic flux in the secondary winding induces AC voltage in the secondary coil.

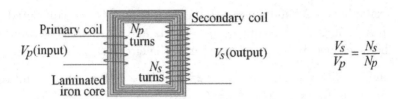

Figure 7.5: Schematic drawing of a transformer that can step up/down AC voltage.

Using Faraday's law of induction, we find the voltage ratio is equal to the turn-ratio of the secondary to the primary windings:

$$\frac{V_s}{V_p} = \frac{N_s}{N_p}, \qquad (7.11)$$

where V_s is the secondary winding electric potential and V_p is the primary AC electric potential. An ideal transformer does not generate or dissipate energy, i.e. $I_p V_p = I_s V_s$. In reality, there is a small resistance in conductors, and there is a small flux dissipation in magnetic materials. Energy dissipation in transformers will become heat.

Transformers are very important in power transmission. Typically, the power companies transport electricity at 110 kV or higher in order to minimize power loss. The electric potential is stepped down to 66 kV and 33 kV at local stations. The electric potential is further stepped down to 120 V in residential homes. Some small home electronics devices use 12 or 24 V. Some early fluorescence lights need high voltage starter, where one needs a transformer to start the gas ionization process.

★ Example 7.11: The number of turns in the primary winding is 300 for a given transformer. What is the number of turns in the secondary to reduce the primary voltage from 120 V to 15 V?

Answer: The turn ratio is 15 V/120 V = 0.125. The number of secondary turns is 0.125 × 300 = 37.5 turns, or 38 turns.

IV. Resonance Circuits and Filters

An electronic circuit with linear electronic elements obeys Ohm's law: $V = I \cdot Z$, where V is the EMF, Z is the resistance or the impedance, and I is the current. Symbols used in electric circuits are DC battery ⊣⊦, AC voltage supply –Ⓐ–, conductor wire ——, resistor –Ⱳ⌐, switch ⟋—, capacitor ⊣⊦⊦, inductor –ⱮⱮ–, and ground ⊥↓, respectively. This section discusses the use of these linear elements to construct resonance circuits, voltage dividers, filters, etc.

A. RLC Circuit and Electrical Resonance

Inductors can store magnetic field energy and capacitors can store electric field energy. A circuit with inductors and capacitors results in an interchange between the magnetic and electric field energies. A circuit with inductors (L) and capacitors (C) is called an LC-circuit. With resistors in the circuit, it is an RLC circuit. Figure 7.6 shows common RLC circuits with the RLC in series (a) and in parallel (b). The resistor R is the "load" that provides the desired functions.

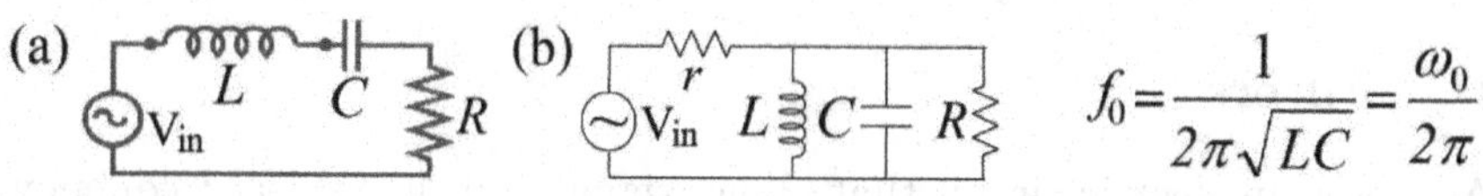

$$f_0 = \frac{1}{2\pi\sqrt{LC}} = \frac{\omega_0}{2\pi}$$

Figure 7.6: RLC circuit in series (a) and in parallel (b). We connected the serial circuit to a voltage source, while the parallel circuit to a current source. The resistor R is the "load."

The impedance of an RLC circuit *in series* is $Z = \sqrt{R^2 + (Z_L - Z_C)^2}$. With an input voltage source V_{in}, the current is $I = V_{in}/Z$. The phase angle φ between the current and voltage, and the power factor of the RLC circuit in *series* in Fig. 7.6(a) are respectively

$$\varphi = \arctan\frac{Z_L - Z_C}{R}, \cos\varphi = \frac{R}{Z} = \frac{R}{\sqrt{R^2 + (Z_L - Z_C)^2}}. \quad (7.12a)$$

The power dissipated in this circuit is $P = IV_{in}\cos(\varphi)$, where $\cos(\varphi)$ is the power factor. It measures the fraction of power delivered to the load in a circuit.

Maximum current or maximum power occurs at the resonant condition: $Z_L - Z_C = 0$, i.e.

$$f_0 = \frac{1}{2\pi\sqrt{LC}}. \quad (7.12b)$$

On resonance, $Z = R$, and the circuit has a maximum current and maximum output voltage. Figure 7.6(b) shows the RLC electric resonant circuit in *parallel*, commonly used in "current AC sources." Its resonance frequency is the same as that of the RLC circuit in series. The voltage across the "shunt" or "load" resistor R is also maximum at the resonance frequency.

★ Example 7.12: Find the resonance frequency of the RLC circuit with (1) $R = 500\ \Omega$, $L = 100$ mH and $C = 0.1\ \mu$F and (2) $R = 1000\ \Omega$, $L = 0.2$ mH and $C = 200$ pF.

Answer: From Eq. (7.12b), resonance frequencies are

$$(1)\ f_0 = \frac{1}{2\pi\sqrt{LC}} = \frac{1}{2\pi\sqrt{(100\times10^{-3}\ \text{H})(0.1\times10^{-6}\ \text{F})}} = 1.6\ \text{kHz},$$

$$(2)\ f_0 = \frac{1}{2\pi\sqrt{LC}} = \frac{1}{2\pi\sqrt{(0.2\times10^{-3}\ \text{H})(200\times10^{-12}\ \text{F})}} = 800\ \text{kHz}.$$

B.   Filters

A filter is an electric circuit that allows signals within a select frequency range to pass through. A low/high-pass filter allows low/high-frequency information in the output signal, while a band-pass filter allows a defined range of frequencies in the output signal. A notch filter blocks a defined range of frequencies in the output signal. Figure 7.7(a) shows a voltage divider circuit with two electronic elements at impedances Z_1 and Z_2 connected in series. The output voltage is

$$V_{\text{out}} = \frac{Z_2}{Z_1 + Z_2} V_{\text{in}}.\tag{7.13}$$

If $Z_2 \ll Z_1$, the output voltage will be small; and if $Z_2 \gg Z_1$, the output voltage will be high. Since the capacitor has a much higher impedance at low frequency [see Eq. (7.10)], we can use an RC circuit for the low-pass filter. Similarly, the impedance of an inductor is small at low frequency; we can also use an LR circuit for the low-pass filter. Reversing the order of these electronic elements in the circuit, we get high-pass filters.

The cutoff frequencies f_{cutoff}, defined as the frequency at 71% of the maximum amplitude or -3 dB reduction in gain function, for the RC

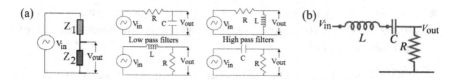

Figure 7.7: (a) Low/high-pass filters using the circuit dividers. (b) An example of band-pass filter using a resonance RLC circuit in voltage divider type circuit.

and RL circuit in Fig. 7.7(a), is respectively,

$$f_{\text{cutoff}} = \frac{1}{2\pi RC} \text{ or } f_{\text{cutoff}} = \frac{R}{2\pi L}. \tag{7.14}$$

Combining the low-pass and high-pass filters, one can design a band-pass filter or band-stop filter (also called notch filter). One can also use an RLC resonance circuit as a band-pass filter, shown in Fig. 7.7(b). At the resonance frequency of Eq. (7.12b), the impedances of the inductor and capacitor cancel each other, and the voltage across the load resistor is maximum. The pass-band frequency is around the resonance frequency of the RLC circuit.

Figure 7.8 shows examples of the first order filters, where (a) is a low-pass filter at cutoff frequency $f_{\text{cutoff}} = 5$ kHz; (b) is the high-pass filter at the cutoff frequency $f_{\text{cutoff}} = 500$ Hz; and (c) band-pass filter by combining the filters (a) and (b). The cutoff frequency is the frequency at the crossing of the -3 dB line in the gain curve. The band-pass frequency is the frequency that has the maximum gain, or $f_c = \sqrt{f_1 f_2}$.

The "first-order filter" has 20 dB/decade slope in the gain function (see dashed lines in Fig. 7.8). The 2nd-order filter is achieved by cascading 2 first-order filters at 40 dB/decade in the gain function, and the cascading process can continue to achieve a 3rd-order filter at 60 dB/decade, and so on.

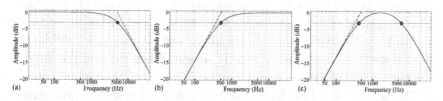

Figure 7.8: First-order filters: (a) low-pass filter, (b) high-pass filter, and (c) band-pass filter.

★ Example 7.13: A resonance circuit has a resonance frequency of $f_0 =$ 440 Hz and a quality-factor Q of 40. What is the linewidth Δf of its resonance curve?

Answer: From Eq. (A.11), we obtain the frequency bandwidth $\Delta f = f_0/Q = 440/40 = 11$ Hz.

★ Example 7.14: Find the RC value for the low-pass filter at $f_c =$ 5 kHz.

Answer: Using the RC circuit for the first low-pass filter shown in Fig. 7.7(a), the cutoff frequency is $f_{\text{cutoff}} = \frac{1}{2\pi RC} = 5000$ Hz from Eq. (7.14), or $RC = 3.2 \times 10^{-5}$ s. One can use combinations of R and C values for achieving this same cutoff frequency, e.g. $C = 100$ pF and $R = 3.2 \times 10^5$ Ω, etc.

V. Active Devices: Amplifier and Operational Amplifier

The circuits discussed so far are passive. They do not contain a power source while power dissipates through all resistive elements. In order to amplify a small signal, we need active devices that provide energy to the circuit. Active devices include vacuum tubes and semiconductor devices.

Active devices can amplify a small input signal. These devices, invented in the beginning of the 20th century, are available in the market, e.g. vacuum tubes, semiconductors, diode, transistors, amplifiers, integrated circuits, operational-amplifier (op-amp), audio-amplifiers, massive integrated circuit, micro-processors, analog-to-digital converter (ADC), digital-to-analog converter (DAC), etc. Figure 7.9(a) shows a schematic drawing of the amplifier principle, wherein a grid or bias voltage can control the output current or voltage. The device, called "grid Audion," was first invented by Lee de Forest in 1906. When the grid potential is negative, the vacuum tube inhibits the flow of electrons from cathode to the anode. When the grid potential is positive, it enhances the electric current. The current vs the grid potential is "nonlinear." This *nonlinearity* can amplify a small input signal. Figure 7.9(b) shows a schematic drawing of a vacuum tube response

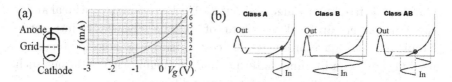

Figure 7.9: (a) Schematic drawing of using the grid potential of a vacuum tube to achieve nonlinearity in the gain curve. (b) Schematic drawing of amplifier classes A, B, and AB, according to its operational region of the biased grid potential.

curve and its use in amplifiers of different classes. Since the invention of the grid Audion, many specially designed vacuum tubes have been developed for various applications.

In 1948, J. Bardeen, W. Brattain, and W. Shockley invented semiconductor triode that had similar function as that of the vacuum tube. Today, semiconductor components are used in modern electronics, except very high voltage and very high power applications. Integrated semiconductor devices are integral part of the voltage amplifiers and power amplifiers that are relevant to acoustics.

The operational amplifiers (op-amp) are linear high-gain high-impedance voltage amplifiers, schematically shown in Fig. 7.10(a). The op-amp symbol is a triangle with inverted $(-)$ and non-inverted $(+)$ input ports. The power supply ports are marked $V^{\pm}$ for positive and negative power supply potential, respectively. An ideal op-amp has infinite open-loop-gain, infinite input-impedance, zero output-impedance, zero off-set voltage, and infinite bandwidth. A typical op-amp has a constant gain-bandwidth-product (GBwP). For example, a voltage-gain of 10^4 in an op-amp with GBwP of 10^6 Hz would have a bandwidth of 100 Hz, or a voltage gain of 100 would have bandwidth above 10^4 Hz.

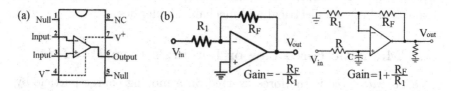

Figure 7.10: (a) Schematic drawing of a popular op-amp 741 that composed of 22 transistors. (b) Possible op-amp circuit connection schemes with inverted input or non-inverted input.

Our audible frequency range is about 2×10^4 Hz, op-amps are ideal audio signal amplifiers up to about a gain of 50. The advances of integrated circuit industry have produced low cost versatile op-amps, extensively used in signal conditioning, filtering, and mathematical operations such as addition, subtraction, integration and differentiation.

Since the op-amps have high gain, we connect op-amps in negative feedback mode, as shown in Fig. 7.10(b). This mode is important in providing stability and high bandwidth in the gain. Connection of input signal through the inverting or non-inverting input ports is shown schematically in Fig. 7.10(b). For example, if the signal input goes through the inverting port, the current flows through input resistor R_1 and then through the feedback resistor R_F. Because the op-amp has very high impedance, there is no current flows through the op-amp. We find $V_{in} \approx IR_1$ and $V_{out} \approx -IR_F$, where I is the current. The voltage gain is Gain $= \frac{V_{out}}{V_{in}} = -\frac{R_F}{R_1}$. The $-$ sign signifies that the output pulse is inverted. When the input signal is connected to the non-inverting $(+)$ input port, the voltage gain is Gain $= 1 + \frac{R_F}{R_1}$. In this circuit example, the RC circuit provides a low-pass filter for the input signal [see also Fig. 7.7(a)].

VI.   Transducers

There are many forms of energies, e.g. mechanical energy, acoustic energy, thermal energy, chemical energy, electrical energy, and electromagnetic wave energy. *Transducers are devices that convert one form of energy to another.* Loudspeakers convert electric signals to acoustic signals and microphones transform acoustic signals into electric signals. The acoustic waves are pressure wave in medium, a form of mechanical energy. The loudspeakers and microphones are transducers that transform between electrical and mechanical energies.

A.   Magnetic Force Law or Lorentz's Law: $F = I\ell B$

The *Lorentz force* is the force exerted on a moving charge particle by the magnetic field. Since electric current is charge moving through conductor wires, the magnetic field will exert a force on the electric current. The magnitude of the Lorentz force on the current in a conducting wire by the magnetic field is $F = I\ell B$. Here F is the force on the wire in [newton], I is the electric current in [ampere], ℓ is the length of the wire in [meter], and B is the component of magnetic flux density (or loosely called magnetic field) in [tesla] or [T] that is perpendicular to the direction of the electric current. The direction of the force is perpendicular to both the current and the magnetic field using the right-hand rule. The Lorentz force law is the main mechanism by which electromagnetic energy can be converted into mechanical energy.

In fact, the unit of electric current, the ampere, is one of seven SI base units in physics. The Lorentz force law is used to define the unit "ampere" as "the constant current which, if maintained in two straight parallel conductors of infinite length, of negligible circular cross section, and placed 1 meter apart in vacuum, would produce a force per unit length between these conductors equal to 2×10^{-7} [N/m]." – https://www.nist.gov/pml/weights-and-measures/si-units-electric-current

B. Loudspeakers

The loudspeaker is a transducer that converts an audio electrical signal into a sound pressure wave. Loudspeakers include electromagnetic transducers, balanced armature transducers, electrostatic or piezoelectric transducers, etc. The electromagnetic force on the electric current in a coil affixed to a diaphragm produces pressure waves at the same frequency as the electric current. The design of loudspeakers with a flat frequency response is a technical marvel. High-fidelity speakers divide the system into tweeters and woofers for enhancing frequency response. Figure 7.11 shows schematic drawings of speaker components and circuit connections of speakers. The electric current at audio frequency runs in the voice coil, in the presence of a magnetic field, inducing mechanical oscillations of the diaphragm at the same frequency (Lorentz's law). The mechanical oscillation transduces the electric current into an acoustic pressure wave.

The main specs for loudspeakers are frequency response and sensitivity. The frequency response is the frequency range, typically 30 Hz to 20 kHz, wherein the response is within ± 3 dB. The speaker sensitivity is measured as the sound intensity level (in dB) at 1 [meter] from the speaker at 1 [watt] power input to the speaker, e.g. [85 dB/W at 1 m].

The top of Fig. 7.11(b) shows a schematic drawing of the "open" load voltage V_L on the load impedance of R_L. The power delivered to the speaker is V_L^2/R_L. Typical speaker has a load impedance $R_L = 8\ \Omega$. In open circuit, the output power is $1\ [\text{W}] = V_L^2/R_L$, so the equivalent voltage is $V_L = \sqrt{8} = 2.83\ [\text{V}]$. Thus, the sensitivity of [85 dB/2.83 V] for 8-Ω speaker is equivalent to the sensitivity [85 dB/W].

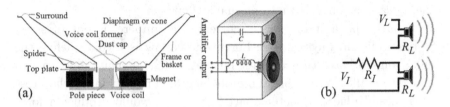

Figure 7.11: (a) Schematic drawing of speaker components. A schematic wiring of a speaker system with low- and high-pass filters to control the cutoff-frequency on the audio-signal for treble or bass speakers. (b) Schematic drawing of a speaker's wiring with input voltage V_I and input impedance R_I connected to a speaker with load impedance R_L (bottom), and an equivalent open circuit (top).

When connecting a speaker to an amplifier, impedance matching is important. Figure 7.11(b) shows the connection of amplifier input impedance R_I and speaker load R_L. Using Ohm's law, the current is $I = V_I/(R_I + R_L)$. The power delivered to the speaker P_{out} and the total power P_{total} are

$$P_{\text{out}} = I^2 R_L = V_I^2 \frac{R_L}{(R_I + R_L)^2} \quad \text{and} \quad P_{\text{total}} = V_I^2 \frac{1}{R_I + R_L}. \qquad (7.15)$$

The maximum power delivered to the speaker P_{out} occurs at the matched load: $R_L = R_I$. The power dissipated in the amplifier is $P_{\text{total}} - P_{\text{out}}$. It is safe to have $R_L > R_I$, at a reduced output power, but if one connect a low impedance speaker to the amplifier with $R_L < R_I$, the increase in current can cause nonlinear response in the power amplifier, and the extra-power dissipated as heat may damage the power amplifier.

★ Example 7.15: We consider an amplifier rated 100 W with 8 Ω impedance. Find the power deliver to the speaker if the impedance of the speaker is (a) 8 Ω; (b) 16 Ω; and (c) 4 Ω.

Answer: First, the amplifier is supposed to drive an 8 Ω speaker for a matched load. The total resistance is $R = 16$ Ω and the rms voltage source is $V_I = \sqrt{RP} = \sqrt{16\,\Omega \times 100\,W} = 40$ V.

(a) At the matched load, the amplifier will deliver 50% of power to the load. Since the amplifier is rated 100 W, it delivers 50 W to the speaker and dissipates 50 W as heat in the amplifier.

(b) Now, if we connect a 16 Ω speaker to the amplifier, the total resistance is $16 + 8 = 24$ Ω, and the current in the circuit is $I = \frac{V_I}{R_I + R_L} = \frac{40}{24} = 1.66$ A. The power delivered to the speaker is $P_{\text{out}} = I^2 R_L = \left(\frac{40}{16+8}\right)^2 \times 16 = 44$ W, and the total power drawn from the amplifier is $P_{\text{total}} = \frac{40^2}{16+8} = 67$ W, i.e. 23 W of power is *dissipated as heat* in the amplifier. The speaker power output is less than 50 W, but also less heat dissipated in the amplifier.

(c) If we connect an expensive 4 Ω speaker to the amplifier, the total resistance is $8 + 4 = 12$ Ω, and the power deliver to the speaker is still $P_{\text{out}} = \left(\frac{40}{4+8}\right)^2 \times 4 = 44$ W. However, the amplifier draws $P_{\text{total}} = 40^2/(4+8) = 133$ W power. In other words, 89 W is dissipated as heat in the amplifier. The power amplifier may have heat load problems and fail.

★ Example 7.16: Show that the speaker sensitivity of 85 [dB/W at 1 m] has an efficiency of 0.4% acoustic energy production.

Answer: The sound intensity at 1 m for 85 dB sound level is $I = 10^{85/10} \times 10^{-12} = 10^{-3.5}$ W/m². The total acoustic power is $4\pi r^2 I = 4\pi \times 1^2 \times 10^{-3.5} = 0.004$ W. The efficiency is therefore 0.4%, i.e. 0.004 W of the acoustic power out of 1 W of electrical input power. A speaker that could convert 1% of electrical energy to acoustic energy would have 89 [dB/W at 1 m].

★ Example 7.17: Loudspeaker sensitivity is measured as the sound intensity level in [dB at 1 m] from a speaker with 1 W of input power supplied to the speaker. Good speakers have 87 [dB/W at 1 m]. What is the efficiency of such a speaker?

Answer: With 1 W power supply, the sound intensity is $I = 10^{87/10} \times 10^{-12} = 5.0 \times 10^{-4}$ W/m^2 at 1 m from the speaker. The power delivered by the speaker is $4\pi r^2 I = 0.0063$ W, where $r = 1$ m. The efficiency is 0.0063 W/1 W = 0.63%.

Headphone, invented in 1910 by Nathaniel Baldwin, is a very convenient device that is very popular today. The frequency response of modern headphone (or earbud) is normally flat from 20 to 20,000 Hz. The sensitivity of earbuds is commonly expressed in dB(SPL)/mW or dB(SPL)/V. Using Ohm's law, we find

$$\frac{\text{dB(SPL)}}{\text{V}} = \frac{\text{dB(SPL)}}{\text{mW}} - 10 \log \left(\frac{R_L}{1000} \right) \tag{7.16}$$

where R_L is the impedance of the headphone, typically 25–600 Ω. The sensitivity of an earbud with 100 dB(SPL)/V can produce 100 dB at the input rms voltage of 1 V.

When some solid-state materials (crystals, certain ceramics and biological matters) are compressed, electric charges accumulate on their surfaces. This phenomenon is the "piezoelectric" effect. Applying an electric field at audio frequency to these materials, the materials will change shape and generate pressure wave at the same frequency. These materials are transducers for converting an electrical wave into a mechanical acoustic wave. Some earphones also use piezoelectric crystal to convert the electric signal into mechanical energy for the acoustic pressure wave.

C. Microphone

A microphone converts the sound (pressure) wave signal into an electric signal. Microphones use various physical mechanisms to convert acoustic pressure waves into electric signals, and they include the piezoelectric crystal microphone, dynamic microphone, condenser (capacitor)

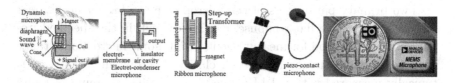

Figure 7.12: Schematic drawing of different types of microphones that convert the acoustic mechanical energy to electric signals. The size of a MEMS microphone vs the size of US dime.

microphone, electret-condenser microphone, ribbon microphone, and MEMS microphone, schematically shown in Fig. 7.12.

A dynamic microphone is the reverse process of the dynamic speaker, which is also usable as a microphone. The capacitance of a capacitor depends on the distance between two parallel plates [see Eq. (7.6)]. The vibration of one condenser plate therefore changes the capacitance and the amount of charge on the plates if the electric potential is kept constant. This is the mechanism used by condenser microphones to convert mechanical vibration into charge motion or to convert a sound pressure wave into electric current. Condenser microphones have a flat frequency response and are commonly used for precision measurements. Similar to the dynamic microphone, ribbon microphones employ Faraday's law of induction to convert mechanical motion into an electrical signal. The piezoelectric crystal can convert pressure into electric voltage or vice versa. It can serve as a transducer for converting acoustic energy into electric energy. The electrets, discovered in the 1920s, are materials that have permanently embedded static electric charge. The development of electret thin film at Bell Labs in 1962 provided new applications for the electret-condenser microphone. Electret-condenser microphones were popular in older PCs; however, the microelectromechanical system (MEMS) microphones, successfully commercialized in 2003, have become the technology of this decade. The MEMS microphone, produced by etching a pressure-sensitive diaphragm directly onto a silicon wafer with integrated preamplifier, has penetrated into mobile phones, hearing aids, headsets, laptops, smartphones, wearable devices, smart home devices, and automotive markets due to recent innovations and its small size of a few mm^3. Both capacitive and piezoelectric MEMS microphones are available. The technology is still evolving.

Two primary parameters of the microphone are *frequency response* and *sensitivity*. The flat response, e.g. ±3 dB within the specified frequency range, is good for measurements, while some microphones tailor their frequency response for speech clarity. Microphone sensitivity is typically measured with a 1-kHz sine wave at a 94-dB sound pressure level (SPL), or 1 pascal [Pa] pressure. The common unit of microphone sensitivity is

$$M_0 = \frac{V_{\text{out}}}{p_{\text{in}}} \ [\text{Volts/Pascal}] \text{ or } [\text{V/Pa}], \tag{7.17a}$$

where p_{in} is the input sound pressure and V_{out} is the open-circuit output voltage. Expressing the sensitivity in [dBV] scale or [dB re 1 V/Pa], we use

$$S_V \ [\text{voltage sensitivity in dBV}] = 20 \log M_0 = 20 \log \frac{V_{\text{out}}}{p_{\text{in}}}. \tag{7.17b}$$

Note that $S_V = 0$ [dBV] at $V_{\text{out}}/p_{\text{in}} = 1$ [V/Pa]. Typically, microphones produce electric potential in millivolt or [mV], one also uses [mV/Pa] as reference, i.e.

$$S_V \ (\text{voltage sensitivity in dBmV}) = 60 + S_V \ (\text{voltage sensitivity in dBV}). \tag{7.17c}$$

We can also express the microphone sensitivity in power output with

$$S_P \ (\text{power sensitivity in } [\text{dBm re 1 mW}]) = 10 \log \frac{W}{1 \, \text{mW/Pa}}. \tag{7.17d}$$

At the microphone power output $W = 1$ mW/Pa, we have $S_p = 0$ dB. The average power in open circuit is W (in mW) $= 1,000 \, V^2/Z$, where V is the voltage of the microphone in [V], and Z is the impedance in [Ω]. The matched output voltage is $V = V_{\text{oc}}/2$, where V_{oc} is the open circuit voltage. Since the power is W (in mW) $= 1,000 \, (V_{\text{oc}}^2/4Z)$, the relation between the S_P and S_V is

$$S_P \ (\text{power sensitivity in } [\text{dBm wrt 1 mW}]) = S_V - 10 \log Z + 24. \tag{7.17e}$$

Here S_V is in Eq. (7.17b). The number 24 comes from $10 \log(1,000/4) = 24$. For a given sound level L_p, the open-circuit output of a microphone is

$$20 \log V = S_V + L_p - 94, \tag{7.18}$$

where 94 dB is the reference sound level of 1 [Pa] rms pressure in the S_V unit.

★ Example 7.18: Microphone sensitivity is typically measured by using a 1-kHz sine wave as a sound source at 94 dB sound pressure level (SPL), or 1 pascal pressure. A microphone exposed to 94 dB SPL generates 2.0 mV of open-circuit output voltage. What is its sensitivity?

Answer: Since 94 dB SPL corresponds to an RMS acoustic pressure of 1 Pa, the microphone sensitivity is 2.0 mV/Pa, or $M_0 = 0.002$ V/Pa, which gives $S_V = 20 \times \log(0.002) = -54$ [dBV/Pa] $= 60 - 54 = 6$ [dBmV/Pa].

★ Example 7.19: A microphone has a sensitivity at 3.5 mV/Pa with an output impedance 200 Ω. What is its sensitivity in M_0, dBV, dBmV, dBm? If you speak to it at 80 dB SPL, what is the open-circuit voltage of the microphone?

Answer: Since $M_0 = 0.0035$ V/Pa, we obtain $S_V = 20\log(0.0035) = -49$ dBV or 11 dBmV. Thus, the power sensitivy is $S_P = -49 - 10\log(200) + 24 = -48$ dBm. At 80 dB SPL applied to the speaker, the open circuit voltage is $20\log(V) = -49 + 80 - 94 = -63$ dB, so $V = 10^{-63/20} = 0.70$ mV. One can also calculate the open-circuit voltage as follows. At 80 dB SPL, the sound rms pressure is $p = 10^{(80/20)} \times 20$ μPa $= 0.2$ [Pa], and the open circuit voltage is $M_0 \times p = 0.0035$ V/Pa $\times 0.2$ Pa $= 0.70$ mV. Both calculations give the same result.

The most recent advance in microphone technology is the MEMS microphone, which uses either capacitive or piezoelectric transducers. One can use M_0 or S_V of Eq. (7.17) to measure the sensitivity of MEMS microphones. However, the MEMS microphone amplifies and digitizes its signal in an application-specific integrated circuit (ASIC). The "sensitivity" of digital microphones use "% of full scale" or DBFS (decibels relative to full-scale), which is the difference between the maximum acoustic signal and the reference SPL at 94 dB "peak." The Acoustic Overload Point (AOP) is the maximum (or full scale) digital microphone's SPL. The sensitivity is

$$S_{\text{digital}} = 20 \log(\% \text{ of } L_p = 94 \text{ dB to full scale } L_p). \qquad (7.19)$$

Microphones are classified into low impedance (dynamic microphone) at 50 to 600 ohms, and high impedance, e.g. crystal, condenser, and electret-condenser, at 50,000 ohms or higher. High frequency signals are highly attenuated in long cables; the loss is typically 6 dB loss per 20 feet of cable for high impedance microphones. High impedance microphones are usually connected to a line transformer or pre-amplifier in the transmission process for impedance match. Most commercial condenser microphones include a build-in transformer resulting in low output impedance. Low impedance microphones can be transmitted through low-impedance cable (600 ohms) without loss of fidelity. Other important issues in sound transduction are dynamic range (DR), signal to noise ratio (SNR), and total harmonic distortion (THD), defined as THD $= (\sum_{i>1} V_i^2)/V_1^2$, where V_i is the signal amplitude of the i-th harmonic and V_1 is the principal signal.

★ Example 7.20: The AOP of a digital microphone is 120 dB, what is its sensitivity?

Answer: The full-scale output (or AOP) is 120 dB. The sensitivity is $S_{\text{digital}} = 94 - 120 = -26$ dBFS. The sound pressure is 1 [Pa] at sound level 94 dB, and 20 [Pa] at 120 dB AOP. The pressure of 94 dB is 5% of the full-scale pressure at 120 dB, we find $S_{\text{digital}} = 20 \log(0.05) = -26$ dBFS. Normally, we measure the digital sensitivity reference to the peak value of the sound wave. If one uses the rms value, the rms pressure of peak 94 dB SPL is $1/\sqrt{2}$ [Pa] or rms $94 - 3 = 91$ dB. The sensitivity of this digital microphone is $91 - 120 = -29$ rms DBFS.

VII. Summary

Electricity and magnetism are important in audio technology. We reviewed some passive linear electronic elements such as resistors, capacitors, and inductors. They obey Ohm's law in direct current (DC) or alternating current (AC) circuits. Using simple voltage divider circuit, one can design various types of filters, e.g. low-pass, high-pass, band-pass and notch filter. In addition, one can design band-pass or notch filters using resonant circuits. All filters have wide applications in audio technology.

Faraday's law of induction and the availability of ferrite materials provide us with the design for a device called a transformer, which can be used to step-up or step-down AC voltages. Faraday's law is also used in transforming mechanical energy to AC-electricity in electric power generators.

The advance of active devices such as the vacuum tube and semi-conductor devices can provide signal amplification, signal manipulation, impedance matching, Analog to Digital Converter, Digital to Analog Converter, etc. We send audio-video signals via wired or wireless transmission by amplitude modulation or frequency modulation. Finally, this chapter discussed the basic principle and characteristics of transducers, including the microphone and loudspeaker.

VIII.   Homework 07

1. In the transmission of electrical energy with alternating currents (AC), transformers are electromagnetic devices used to step up or step down the AC _____ .

2. A 100-W personal computer (PC) runs for 365 days. If the electric company charges you $0.10/kWh for electric energy, the cost of electricity for running the PC is $_____ .

3. A simple circuit example is shown at the left plot in Fig. 7.1. What is the current of the circuit? If the circuit runs for 1 hour, how much charge runs through the circuit?

4. Three light bulbs with resistance 20 ohms each are connected in parallel to a power source that generates 12 V voltage. The current through one of the light bulb is _____ amps and the power of each light bulb is _____ W.

5. Three light bulbs are connected in series to a power source at 12 V. Each bulb has a resistance of 20 ohms. The current passing through each light bulb is _____ amps and the power delivered by each light bulb is _____ W.

6. A 50 W incandescent light bulb is connected to an AC line at rms voltage 120 V. What is the resistance of the light bulb?

7. Two 50 W incandescent light bulbs are connected in series to rms 120 V AC line. What is the power delivered to each light bulb?

8. The rms current in an ac circuit with a resistance of 100 Ω is 0.86 A. The average power of this circuit is _____ W, and the maximum power consumed by this circuit is _____ W.

9. The reactance of a capacitor is 55 Ω at a frequency of 1000 Hz. The capacitance of the capacitor is _____ $\times 10^{-6}$ F.

10. A 12.0×10^{-6} F capacitor is connected to an ac generator with an *rms* voltage of 24.0 V and frequency of 180 Hz. The rms current in this circuit is _____ A. The power dissipated in the capacitor is _____ W.

11. A 100 mH inductor is connected to an ac generator with an *rms* voltage of 24.0 V and frequency of 180 Hz. The rms current in this circuit is _____ A.

12. An RLC circuit with R = 1 kΩ, L = 10 mH, and C = 10 μF are connected in series. What is its resonance frequency?

13. At what frequency will a generator with an rms voltage of 120 V produce an rms current of 10.0 mA in a 20×10^{-9} F capacitor? _____ Hz.

14. One vibrating system has a resonance frequency of 440 Hz and a quality-factor Q of 40. The linewidth Δf of its resonance curve = _____ Hz.

15. A neon sign that requires a voltage of 2.4 kV is plugged into a 120-V wall outlet. What turns ratio (secondary/primary) must a transformer have to power the sign? The turn ratio of secondary to primary is _____ .

16. Find the power factor for an RC circuit connected to a 60 Hz generator with an rms voltage of 120 V. The values of R and C in this circuit are 200 Ω and 5.0×10^{-6} F, respectively. The power factor is _____ .

17. Find the RC time constant value for the low-pass filter at $f_{cutoff} =$ 10 kHz.

18. Find the RC time constant value for the high-pass filter at $f_{cutoff} =$ 500 Hz.

19. Find the L/R time constant value for the low-pass filter at $f_{cutoff} =$ 5 kHz.

20. Find the L/R time constant value for the low-pass filter at $f_{cutoff} =$ 500 Hz.

21. The first-order filter has a _____ dB/decade decrease in the gain function. The 2nd-order filter has a _____ dB/decade decrease in the gain function.

22. An amplifier rated is rated at 25 W with 8 Ω impedance. What is the power deliver to a 4 Ω speaker connected to the amplifier?

23. A loudspeaker rated at 85 dB/W at 1 m are considered to be good. What is the efficiency of the speaker? What is the speaker output power if a power amplifier of 50 [W] is connected to this speaker at the matched load?

24. Microphone sensitivity is typically measured at frequency of _____ Hz sine wave at a _____ dB sound pressure level (SPL).

25. A microphone sensitivity is $S_P = -54$ dBm and a 500 Ω output impedance. What is its voltage sensitivity and open circuit voltage output of a singer singing at 100 dB?

26. What is the microphone sensitivity in S_V for a 50 Ω microphone rated at open circuit 4.0 mV/Pa?

Chapter 8

Digital Technology in Acoustics

In 1876, Graham Bell invented the telephone by transduction of sound pressure waves into electrical signals and converted the electrical signals back to pressure waves via analog transduction. In 1943, the SIGSALY secure speech system provided digitalized voice signal transmission by using the pulse code modulation (PCM) patented by Alec Harley Reeves in 1938. Modern MEMS microphones employ integrated circuit to produce digitized output signal.

In the 20th century, electronic computing technology took off propelled by the advance of semiconductor technology. Voice synthesis became available in personal computers, smart phones, smart devices, etc. Speech recognition programs routinely served commercial communications and transactions between human and various computer systems. The artificial intelligence (AI) in humanoid can synthesize human speech and understand basic human interaction. This chapter explores the basic digital technologies in acoustics, including digitization, the Nyquist theorem, dynamic range, CD technology, etc. Digitization is the trend of world information. Binary information is stored in various storage devices. Digital storage grew from 1% in 1986, 3% in 1993, 25% in the year 2000; and 94% ~ 97% in 2007 at about 2.95×10^{20} bytes.[1]

---

[1]See Martin Hilbert and Priscila López, *Science*, **332**, 60 (2011); DOI: 10.1126/science.1200970

I.   Digital Computer Fundamentals

Number and counting are integral parts of human cultures. Historically, there were a few number systems among human societies, based on the basic unit of 2, 4, 8, 10, 12, 16, etc. The "decimal" system created by Indian-Arabic civilization became the prevailing number system in the world. The memory of digital computers uses 0/1 or on/off states for the binary system. We list these number systems as follows:

(1) Decimal system: A decimal system, commonly used in daily life, bases on 10. For example, the number 10518 stands for $(10518)_{10} = 1 \times 10^4 + 0 \times 10^3 + 5 \times 10^2 + 1 \times 10^1 + 8 \times 10^0$.

(2) Base-8 and base-16 number system: The base-8 (octal) and base-16 (hexadecimal) number systems had been used in many societies. For example, the weight measure 1 lb is 16 oz., and the 2:5 beads abacus can function as a base-16 system. For example, number $(10518)_{10}$ are

$$(10518)_{10} = (24426)_8 = 2 \times 8^4 + 4 \times 8^3 + 4 \times 8^2 + 2 \times 8^1 + 6 \times 8^0$$

$$(10518)_{10} = (2916)_{16} = 2 \times 16^3 + 9 \times 16^2 + 1 \times 16^1 + 6 \times 16^0$$

The digits for the base-8 system are 0, 1, 2, 3, 4, 5, 6, 7; and digits for the base-16 system are 0, 1, 2, 3, 4, 5, 6, 7, 8, 9, A, B, C, D, E, F, e.g. $(1F)_{16} \equiv 1 \times 16^1 + 15 \times 16^0 = (31)_{10}$.

(3) Binary system: The computer memory use binary system because an electronic circuit can easily manipulate on/off states. The state "on" is 1 and "off" 0. The number system is binary algebra. The unit of 1 or 0 is called a "*bit.*" A 2-bit system represents 00, 01, 10, and 11. There are 4 possible combinations. An 8-bit system has 256 combinations, and N-bits system has 2^N combinations. For example, any integer number lying in [0, 255] can be represented by the 8-bit binary number system, e.g. 65 in decimal is $65_{10} = 0 \times 2^7 + 1 \times 2^6 + 0 \times 2^5 + 0 \times 2^4 + 0 \times 2^3 + 0 \times 2^2 + 0 \times 2^1 + 1 \times 2^0 = (01000001)_2$. Similarly, $(10518)_{10} = (10100100010110)_2$. Here the first binary digit is the most significant bit (MSB), and the last binary digit is the least significant bit (LSB). There are two types of numbers storage in digital computer systems: integers and floating-point numbers.

1. Characters, words and integers: A standard unit in computers is the 8-bit unit known as the *byte*. A byte can be represented 00, 01, up to FF in hexadecimal. There are many character-coding sets for world languages. ASCII code uses 1 byte for 128 possible English characters. Unicode is a computing industry standard for encoding, representing, and handling of text. The UTF-8 is "variable width" character, encoding up to 4 bytes. Today, UTF-8 is the prevailing encoding for the World Wide Web and is backward compatible to the ASCII code. The storage of a word is 2 bytes. The storage of an integer number varies from 4 bytes, 8 bytes, up to 16 bytes.

2. Floating-point numbers: One uses scientific notation to represent decimal numbers. Scientific notation is $[\text{sign}][\text{mantissa}] \times 10^{[\text{exponent}]}$, where the mantissa is the significand or coefficient. For example, we represent $+0.123$ with $[+][123] \times 10^{-3}$. Computers and microprocessors store floating-point numbers in binary form: $[\text{sign bit}][\text{mantissa}] \times 2^{[\text{exponent}]}$. The number of bits reserved for the mantissa and exponent depends on the precision, set by the IEEE-754 standard (see https://standards.ieee.org/standard/754-1985.html for details and history) and ISO/IEC/IEEE 60559:2011 Floating point arithmetic. Since 2008, most processors for scientific applications have floating point math co-processor with transcendental functions.

★ Example 8.1: Convert the following decimal numbers, 12, 22, 65, and 127 to binary numbers.

Answer: The procedure (a) to (e) listed below converts the decimal number 37 to a binary number. The general steps are: (1) successively divide 37 by 2 and mark the remainders on the right, and (2) collect all remainder to find the binary representation of 37.

```
2 |37_ |1      2 |37_ |1      2 |37_ |1      2 |37_ |1      2 |37_ |1
   18        ⇒  2 |18_ |0   ⇒  2 |18_ |0   ⇒  2 |18_ |0   ⇒  2 |18_ |0
                   9             2 |9_ |1      2 |9_ |1      2 |9_ |1
                                    4          2 |4_ |0      2 |4_ |0
                                                  2          2 |2_ |0
                                                                1        37₁₀= 100101₂
    (a)           (b)            (c)            (d)            (e)
```

$37_{10} = 100101_2$

Using the above procedure, we can find $12 = (1100)_2$, $22 = (10110)_2$, $65 = (1000001)_2$, and $127 = (1111111)_2$. To convert a decimal number to base-4 system, we divide it by 4 and carry out similar procedure.

★ Example 8.2: Convert the binary numbers, 0110, 00101101, 10000001, and 1111111111111111 to decimal numbers:

Answer: Use the definition of the binary number, we find $0110 = 0 \times 2^3 + 1 \times 2^2 + 1 \times 2^1 + 0 \times 2^0 = 4 + 2 = 6$. Similarly, one finds $(00101101)_2 = 45$, $(10000001)_2 = 129$, and $(1111111111111111)_2 = 65535$.

II.　Signal Digitization

A common digitization method for a sound wave is pulse code modulation (PCM), which digitizes an analog signal at a given *sampling time step* Δt_s and *sampling depth*. The sampling rate $R = f_s$ (also called the sampling bandwidth) is the number of samples per second. The sampling time step is $\Delta t_s = 1/f_s$. The sampling depth is the number of bits used to represent the signal strength. The PCM records an analog signal at an uniform time intervals Δt_s. The Linear PCM (or LPCM) divides the peak-to-peak amplitude into equal interval and quantifies the magnitude of the wave to the nearest value at the sampling depth. For example, if we wish to digitize a signal from 0 to 25 V with 4-bit sampling depth digitizer, we set the reference voltage to $V_{\text{ref}} = 25$ V, and we divide the signal by the reference voltage so that the signal is between 0 and 1, i.e. we represent the voltage signal as

$$\frac{V}{V_{\text{ref}}} \Longleftrightarrow a_1 2^{-1} + a_2 2^{-2} + a_3 2^{-3} + a_4 2^{-4} \tag{8.1}$$

where a_1, a_2, a_3, and a_4 are either be 0 or 1, where a_1 is the most significant bit. The sampled data is $a_1 a_2 a_3 a_4$. The Table below lists the 4-bit number vs the decimal. For example, for a signal at $V = 15$ V,

Table 8.1: 4-bit binary representation of a real number between 0 and 1 of Eq. (8.1).

1111	0.9375	1011	0.6875	0111	0.4375	0011	0.1875
1110	0.875	1010	0.625	0110	0.375	0010	0.125
1101	0.8125	1001	0.5625	0101	0.3125	0001	0.0625
1100	0.75	1000	0.5	0100	0.25	0000	0

we find $V/V_{\text{ref}} = 0.60$ that lies between 0.5625 and 0.625. This voltage is represented by $1001 = 1 \times 2^{-1} + 0 \times 2^{-2} + 0 \times 2^{-3} + 1 \times 2^{-4}$ for 4-bit digitization. In fact, the digitized number 1001 represents any signal lies in [14.0625 V, 16.625 V). Similarly, 0100 represents a signal at $0.25 \leq V/V_{\text{ref}} < 0.3125$, as listed in Table 8.1.

Figure 8.1(a) shows a LPCM digitized sine wave digitized at the sampling depth of 4 bits. We sample the sine wave at regular time interval, shown as vertical lines. For each sample, the data value (on the y-axis) is assigned according to a number depending on the bit-depth. At 4-bit-depth, the sampled value takes the quantized value listed in the Table 8.1. Figure 8.1(b) shows a schematic PCM digitization at infinite bit-depth. The dashed line is the original analog signal.

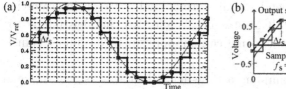 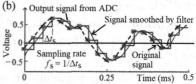

Figure 8.1: (a) A schematic drawing of PCM digitization of a sine signal (thin line) with 4-bit digitization (thick line). (b) A schematic drawing of sampling an arbitrary waveform.

If we have more memory in the digitizing system, we can sample the data in 8-bit sampling depth, and represent the data in $a_1a_2a_3a_4a_5a_6a_7a_8$. The digitized data will have finer resolution at the expenses more storage memory. Typical digitizing systems use 16-bit sampling depth. Some high-resolution audio systems use 20 to 24 bit sampling depth.

★ Example 8.3: Suppose you have an electronic signal with possible voltage values "V" between 0 and 16 volts, and you want to represent these values with a 4-bit number ranging from smallest (0000, for 0 volts $\leq V < 1$ volt) to largest (1111, for 15 volts $\leq V \leq 16$ volts). What is the bit pattern for the voltage value $V = 9.5$ volts?

Answer: 1001 (Since the binary representation of 9 is 1001, 9.5 V is represented by 100.)

★ Example 8.4: Suppose you have an electronic signal with possible voltage values "V" between 0 and 1 volts, what is the bit pattern for voltage value at $V = 0.59$ volts with a 4-bit digitizer?

Answer: 1001 (Since $0.59 \in [0.5625, 0.625)$, it is digitized to 1001. See also Table 8.1.)

A.   Sampling Depth and Sampling Rate

Sampling rate and sampling depth are the main characteristics of signal digitization. The sampling depth is the number of bits used to represent each data point. Common sample depths in LPCM are 8, 16, 20 or 24 bits per sample. There is also a 32-bit LPCM system with support of sound cards. An LPCM encodes a single sound channel. Support for multichannel audio depends on file format and relies on interweaving or synchronization of LPCM streams. While two channels (stereo) is the most common format, some can support up to 8 audio channels (7.1 surround).

In the LPCM, we sampling a time-dependent wave signal in equal time-interval Δt_s. The sampling rate is $f_s = 1/\Delta t_s$. For example, the sampling rate is 1 kHz if we digitize a wave at every 1 ms time interval, and the sampling rate is 1 MHz if we sample a wave at every 1 μs time interval. Commonly used sampling rates are 44.1 kHz for Compact discs (CD), 48 kHz for DVD format video, and 96 kHz and 192 kHz for newer equipments. The CD-Digital Audio standard uses 2-channel 16-bit depth Linear PCM at sampling rate 44,100 Hz. This format requires $2 \times 44.1\,\text{kHz} \times 16\,\text{bits} = 1.41\,\text{Mbits/s}$, or 84.7 Mbits/min of data memory. The DVD format videos use sampling rate of 48 kHz and up to 8-channels. The bit-rate is (8 channels) $\times$ (48 kHz) $\times$ (16-bit per sample) $= 6.144$ Mbits/s for LPCM audio on DVD-Video. DVD audio and Blue-ray Disc can support up to 24 bits per sample.

B.   Dynamic Range (DR)

Sound waves in air, electrical signals in electronic circuits, and our electrical neuron pulses are analog signals. We can easily amplify, filter, and reproduce analog signals by using electronic devices. However, an analog signal is susceptible to noise and thus has relatively small dynamic range

(DR), defined in decibel scale as

$$DR = 20 \log \frac{V_{\text{amplitude}}}{\Delta V}, \qquad (8.2)$$

where $V_{\text{amplitude}}$ is the amplitude of the sound wave and ΔV is the resolvable difference in the wave amplitude. Typical analog electronic have DR $\sim$ 40–50 dB due to thermal noise. The DR is about 40 dB in early 78 rpm phonograph discs, 55–65 dB in Vinyl microgroove phonograph records, and up to 70 dB in high fidelity players. Furthermore, an analog signal gets noisier with each reproduction, while digital data reproduction will retain the fidelity and DR of its original file. The digital audio system of D-bit depth has a dynamic range

$$DR = 20 \log \frac{2^D}{2} = 6.02 \times (D - 1) \text{ dB}. \qquad (8.3)$$

Bit-depth (D)	8	12	16	20	24
Dynamic range (dB)	$48 - 6 = 42$	68	90	114	138

Compact disc (CD) format employs digital electronics to achieve high fidelity, low noise, and great dynamic range at 16-bit depth. Super audio CD, DVD-audio, and Blue-ray audio use higher resolution with 20 or 24-bit depth. For human ears, the jnd of sound level is about 1 dB, the maximum tolerable level is about 130 dB, and the DR is about 100 dB (see Homework 8.1).

★ Example 8.5: In electronic, typical resolvable amplitude is 50 μV due to thermal noise. What is the DR of (a) a DC signal of 0–5 V, and (b) an AC signal with the peak-to-peak wave amplitude 5 Volts?

Answer:

(a) For the DC signal, DR $= 20 \log(5 \text{ V}/50 \,\mu\text{V}) = 20 \log(10^5) = 100$ dB.

(b) The amplitude of AC signal with peak-to-peak voltage of 5 V is 2.5 V. The actual DR of the AC signal is DR $= 20 \log \frac{2.5 \text{ V}}{50 \,\mu\text{V}} = 100 - 6 = 94$ dB, agreeing with Eq. (8.3).

★ Example 8.6: What is the DR of an 18-bit digital representation of an analog signal with amplitude of 10 V (i.e., signal can vary from $+10\,$V to $-10\,$V)?

Answer: Digitization with D $=$ 18 bits would allow for a huge dynamic range of amplitudes, from the smallest binary number 0 to the largest binary number of $2^D - 1 = 262{,}143$. The amplitude resolution (step-size) is $(20\,\text{V})/(2^D - 1) = 7.63 \times 10^{-5}\,$V. Therefore, DR $= 20\log(10\,\text{V}/7.63 \times 10^{-5}\,\text{V}) = 102.3$ dB, which can also be calculated $6.02 \times (18 - 1) = 102.3$ dB. A higher number of bit-depth will have a better resolution, or a larger dynamic range.

III. Sampling Rate and Nyquist Sampling Theorem

The analog-digital-converter (ADC) can convert an arbitrary waveform into a digitized waveform. The digitized waveform depends on the digitizing time step Δt_s. The sampling rate or the digitization bandwidth is $R = f_s = 1/\Delta t_\text{s}$. The Nyquist sampling theorem states that digitization at R samples per second can at best represent a signal containing frequencies up to $\frac{R}{2}$ Hz. The frequency $f_\text{digitized}$ of the digitized signal is

$$f_\text{digitized} = \left| f - \text{int}\left(\frac{f}{f_s} + \frac{1}{2}\right) f_\text{s} \right| \le \frac{R}{2}, \qquad (8.4)$$

where f is the frequency of the original wave, f_s is the sampling rate, the function $\text{int}(x)$ is the integer function that returns the integer lower than the real number x in the argument. The Nyquist frequency limit is $R/2$. Figure 8.2(a) shows the observed frequency of a digitized signal $f_\text{digitized}$ vs the signal frequency f. The spectrum will contain the "aliasing frequencies" folding back from frequencies higher than $R/2$ at their corresponding amplitudes. A low-pass filter before digitization can minimize aliasing signal contamination.

Aliasing is a phenomenon that the frequency of the digitized signal is lower than that of the actual original signal. Aliasing in the digitized data occurs when the sampling rate is smaller than twice the maximum frequency of the original signal. Reconstruction of a signal is possible when the sampling frequency is greater than twice the maximum frequency of the signal. Figure 8.2(b) shows two examples of pulse code digitization at infinite bit-depth. The digitized wave has different period when the sampling rate is less than twice the signal frequency.

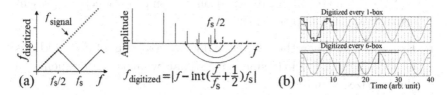

Figure 8.2: (a) Schematic drawing of the Nyquist sampling theorem: The observed frequency of a signal is always lower than $\frac{1}{2}$ of the sampling bandwidth. (b) A sine wave with a period of 8 time units is sampled at $\Delta t_s = 1$ time-unit (top) and 6 time-unit (bottom). The frequency of the sampled wave at $\Delta t_s = 1$ time-unit is the same as the original. The frequency of the sampled wave at $\Delta t_s = 6$ time-units exhibits aliasing effect with a period of 24 time-units.

Typical sampling rates of our digital technology are 8 kHz for telephone; 16 kHz for speech systems; 22.05 kHz in modern games; 44.1 kHz for the CD audio technology and 48 kHz for the DAT (digital audiotape). Since human auditory hearing range is 20 Hz to 20 kHz, digitizing system bandwidth must satisfy the condition $f_s \geq 2 \times 20$ kHz $= 40$ kHz. The Compact Disc (CD) chooses 44.1 kHz. The CD format uses 44.1 kHz sampling rate. The maximum frequency that a CD waveform can represent is 22.05 kHz. The digitized telephone system uses $R = 8$ kHz, which provides a maximum frequency up to 4 kHz. This is good enough for the first two to three formants of our speech. To remove undesirable aliasing in a digitized signal, one can pass the original signal through a low-pass filters that suppresses energy with frequencies higher than the Nyquist frequency $f_s/2$. Besides aliasing, other errors arising from digitization include quantization error due to finite bit-depth (limited dynamic range), phase distortion, etc. Many of these problems can be solved by oversampling, dithering, and companding (simultaneous compression/expansion scheme, patented by A.B. Clark at AT&T, 1928), etc.

★ Example 8.7: When a telephone digitizing system of $f_s = 8$ kHz is used to digitize a 3-kHz sine wave signal, what is the frequency of the digitized signal?

Answer: The digitized wave frequency is $f_{\text{digitized}} = |3 - \text{int}(\frac{3}{8} + \frac{1}{2}) \times 8| = |3 - 0| = 3$ kHz. Note that we use the integer function int $(7/8) = 0$. The telephone digitization system can correctly represent signal frequencies up to 4 kHz.

★ Example 8.8: When the telephone digitizing system of $f_s = 8$ kHz is used to digitize a 4 kHz sine wave signal, what is the frequency of the digitized signal?

Answer: The digitized signal frequency is $f_{\text{digitized}} = |4 - \text{int}(\frac{4}{8} + \frac{1}{2}) \times 8| = |4 - 8| = 4$ kHz. Note that we use the integer function int $(8/8) = 1$.

★ Example 8.9: When the telephone digitizing system of $f_s = 8$ kHz is used to digitize a 6 kHz sine wave signal, what is the frequency of the digitized signal?

Answer: The digitized signal frequency is $f_{\text{digitized}} = |6 - \text{int}(\frac{6}{8} + \frac{1}{2}) \times 8| = |6 - 8| = 2$ kHz, where we use the integer function int $(1.25) = 1$. Similarly, if you use the telephone digitizing system to digitize a 5 kHz signal, you get 3 kHz; digitization of 7 kHz signal gets 1 kHz. If you use the telephone digitizing system to digitize an 8 kHz sine wave, you will get 0 Hz, a DC-signal. If one tries to digitized a signal composed of components with frequencies 3, 5, 6, and 8 kHz, the frequencies of the digitized signal become 3, 3, and 2 kHz with a DC offset.

★ Example 8.10: The sampling rate of digitized telephone and encrypted walkie-talkie, wireless intercom and wireless microphone transmission is 8 kHz. What is the limiting Nyquist frequency of the signal presentation?

Answer: The limiting Nyquist frequency is $(8 \text{ kHz})/2 = 4$ kHz. Any signal with frequency higher than 4 kHz will result in aliasing. To minimize aliasing effect, the telephone signal is passed through a low-pass filter before digitization.

★ Example 8.11: Consider a pure cosine wave signal with period 8 time-unit, as in Fig. 8.2(b). If each time unit (time-slice) is 1 ms, the frequency of the cosine wave is $f = 125$ Hz. If you digitize once every 6 time-units, what is the period of the digitized wave?

Answer: Since $\Delta t_s = 6$ ms, the sampling rate is $f_s = \frac{1}{\Delta t_s} = 166.7$ Hz. The frequency of the digitized wave is $f_{\text{digitized}} = |125 - \text{int}(\frac{125}{166.7} + \frac{1}{2}) \times 166.7| = |125 - 166.7| = 41.7$ Hz, instead of 125 Hz. The bottom plot of Fig. 8.2(b) shows the digitized waveform. Aliasing occurs because the sampling rate of 166.7 Hz is less than twice the signal frequency $2f = 250$ Hz.

IV. Lossless and Non-PCM Lossy Compression Formats of Audio Recording

Popular wave digitization formats are lossless wave-format and the lossy compressed mp3-format. Waveform Audio File Format (WAVE, or .WAV as filename extension) is a Microsoft and IBM audio file format standard (stereo, 16 bits, 44.1 kHz) for storing an audio bitstream (LPCM) on computers. It is an example of a lossless digital format.

A standard CD has a diameter of 120 mm (4.7 in) and can hold up to 80 minutes of uncompressed audio or 700–800 MB of data using 780 nm solid-state laser. The diffraction limit of the 780 nm laser is about $a \cdot \sin(\Delta\theta) = 1.22\lambda$, i.e. the spot size is typically about $a \sim 1.6$ μm, which is also the distance between tracks of a CD. The CD storage is diffraction limited. DVD and HD-DVD or Blue-ray use 650 nm and 405 nm solid-state lasers, respectively. The amount of data storage is proportional to the $1/\lambda^2$ due to the diffraction limit of laser light. The spot size depends also on the "numerical aperture (NA)," where NA $\approx \sin(\Delta\theta)$. Alternatively, NA $\approx 1/(2 \cdot \text{f-number})$, where f-number $\equiv f/d$, f is the focal length and d is the aperture of the laser optical system.

The quantization of the lossy compression formats is performed on PCM samples in the frequency domain, where much of the digitized signals in PCM digitized signals may not be relevant to our listening due to psychoacoustic masking of our hearing system. There is no bit-depth in the frequency domain. Instead, wave frequency data are digitized by 16 to 320 kbits/s. MP3 is currently the de facto standard of digital audio compression for the transfer and playback of music on digital audio players. MP3 stands for "MPEG-1 and MPEG-2 Audio Layer III," developed by the Moving Picture Experts Group (MPEG). The song "Tom's Diner" written in 1981 by American singer-songwriter Susanne Vega was used in the design of the codec (coding/decoding data stream), among audio engineers. This anecdote has earned Vega the informal title "The Mother of the MP3." MP3, patented by Fraunhofer in 1991, is an example of a lossy digital format.

★ Example 8.12: The standard CD has 120-mm diameter with a 15-mm diameter center hole. The track goes from 46 to 117 mm in diameter with track pitch 1.6 μm. There are about 22,188 tracks with a total track length 5.7 km. The area to store information on CD is 90 cm^2. What is the area occupied in each bit?

Answer: The .wav format uses $2 \times 44.1\,\text{kHz} \times 16\,\text{bits} = 1.41$ Mbits/s. If the CD has a playing time of 67 min., the total memory is $1.41\,\text{M} \times 67 \times 60 = 5.67$ Gbits $= 708$ Mbyte stored on the disc. The area for each bit information is 90 cm^2/5.67 Gbit $= 1.60 \times 10^{-12}$ m^2/bit or 1.60 μm^2/bit.

★ Example 8.13: The DVD uses a green laser ($\lambda = 635$ nm). The DVD is to be compared with the CD, which uses an infrared laser ($\lambda = 790$ nm). If the storage capacity is inversely proportional to the square of the laser wavelength, this change in laser wavelength gives the DVD a factor of $(790/635)^2 = 1.55$ more storage capacity than the CD. If a blue laser ($\lambda = 430$ nm) could be used, this factor would increase to $(790/430)^2 = 3.38$.

★ Example 8.14: Find the compression factor from the LPCM in CD format to 128 kbits/s MP3.

Answer: The number of bits per second for LPCM in CD mode is (2-channel) $\times$ (16 bits/channel) $\times$ (44,100 samples/s) $= 1,411,200$ bits/s. Thus the compression ratio is $1,411,200/128,000 = 11.025$, i.e. an MP3 file with 128 kbit/s reduces an audio .wav CD file by a factor of 11, or 7.7 Mbits/min or about 1 MB/min. Similarly, 192 kbit/s and 64 kbit/s MP3 sampling rates have compression ratios of 7.4 and 22 respectively. Typical commercial mp3 compression is higher than 192 kbit/s for music audio files, and it is 64 kbit/s for audiobooks and podcasts.

As the compression increases, the resulting file size decreases. Figure 8.3 compares spectra of various degree of compression, where the left plot spectrum of the Tom's diner sung by Susanne Vega in the uncompressed CD LPCM format. The other spectra are those of mp3 compression at 192 kbps, 128 kbps, and 96 kbps. At 128 kbps, mp3 format is acceptable, because few people can hear sound beyond 15 kHz. Commercial MP3 music tracks use 192–320 kbps.

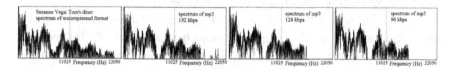

Figure 8.3: Frequency content of LPCM in Susanne Vega Tom's diner recording and the spectra of various mp3 compression at different bandwidths. Note that the reduced high frequency content at higher degree of compression.

Other formats include AAC (advanced audio coding developed by AT&T Bell Labs, Dolby, Sony, and Nokia in 1997), WMA (Microsoft), Ogg Vorbis, etc. Ogg Vorbis is a non-proprietary, patent-and-royalty-free, general-purpose compressed audio format for mid to high quality (8–48 kHz, 16+ bit, polyphonic) audio and music at fixed and variable bitrates from 16 to 128 kbps/channel.

V. Summary

Digitization is the trend of our society in all front of technologies. This chapter discussed the basic binary number system in computer storage and operation; the method of analog-to-digital conversion (ADC), linear pulse code modulation (LPCM) in CD and DVD format, sampling rate, sampling depth, and the Nyquist sampling theorem. To avoid an aliasing problem, the sampling rate must be larger than twice the maximum frequency of the original analog signal.

LPCM files require large digital storage space. The advancement of digital memory and data compression technology pave market growth.

All media have been moving in the direction of the digital revolution. A de factor compression scheme by employing the psychoacoustic frequency masking effect of our hearing system is the MP3 format in data compression. The digitized MP3 digital file has become the standard format in music and audio book market place.

The advantage of digital system is its high dynamic range and its reproducibility in transmission and reproduction. However, we need to convert the digital data to analog signal for various applications. Digital-to-analog converters (DACs) produce an analog voltage or current output that represents digital input data. "Demodulators" that produce accurate analog signals from discrete data are similar to those used for generating the digital signal in reverse.

In the movie "*2001: A Space Odyssey*" (1968), the computer HAL 9000 was "capable of speech, speech recognition, facial recognition, natural language processing, lip reading, art appreciation, interpreting and reproducing emotional behavior, reasoning, and playing chess" (Wikipedia). The inciting incident occurred when HAL tried to take over control of the spaceship. The processor speed then was 100 MHz at best, while a typical microprocessor is now running at the cycle rate of GHz. We are annoyed by the robot calls from the voice synthesizers. Imaging that someday these calls, including video images, may sound and look like our friends and families powered by AI-systems.

VI.   Homework 08

1. The intensity limen (jnd) is about 1.0 dB at sound level 30 dB, and 0.5 dB at sound level 80 dB, the intensity limens are $\Delta I =$ _____ W/m^2 at 30 dB, and $\Delta I =$ _____ W/m^2 at 80 dB. (See also Homework 3.25 in Chapter 3.) The intensity limen depends on sound level. The dynamic range need to consider this variable resolution. If we think that the jnd as the digitization step of hearing system, we use an efficient nonlinear PCM system, i.e. our intensity limen increases with intensity.

2. Compact discs employ _____ electronics to achieve high fidelity, low noise, and great _____ range.

3. A .wav file is an example of a _____ digital format, while a MP3 file is an example of a _____ digital format where not all of the original sound information is present.

4. Compression of a CD standard digitized sound file into a MP3 at 128 kbit/s decreases the file size by a factor of _____ .

5. The _____ theorem states that at least R samples per second are needed to adequately represent a signal containing frequencies up to R/2 Hz.

6. In a digital signal, _____ is the misrepresentation of frequencies _____ (above/below) half the sampling rate at incorrect frequencies _____ (above/below) half the sampling rate. A low-pass filter can minimize this effect.

7. A CD format uses _____ channels with _____ bits at the sampling rate of _____ kHz.

8. The value in binary representation of the following decimal numbers is: $12 =$ _____ , $22 =$ _____ , $65 =$ _____ , and $127 =$ _____ .

9. The decimal value of each of the following binary numbers is: $0110 =$ _____ , $00101101 =$ _____ , $10000001 =$ _____ , and $1111111111111111 =$ _____ .

10. Suppose an 18-bit digital representation of an analog signal with amplitude of 10 V (i.e., signal can vary from $+10\,V$ to $-10\,V$) is to be used. N = 18 bits would allow for a huge dynamic range of amplitudes, from the smallest binary number ($2^0 =$ _____) to the largest binary number of ($2^N - 1 =$ _____). This corresponds to an amplitude resolution (step-size) of $\Delta V = (20\,V)/(2^N - 1) =$ _____ V. The dynamic range is $20\log(10V/\Delta V) =$ _____ dB.

11. For accurate reproduction of audio signals up to 20 kHz, a CD stores 44,100 samples per second for each of the two stereo tracks,

with each sample represented by a 16-bit binary "word." The number of bits per second that must be read by the CD player = _____ bits/s. If the CD has a playing time of 67 minutes, this corresponds to a total of _____ bits being stored on the disc.

12. If the information on the CD is stored in a total area of 80 cm^2, this corresponds to an area for each bit of _____ m^2/bit or _____ μm^2/bit. [This is smaller than the 2.7 μm^2 laser spot size.]

13. The DVD uses a green laser (λ = 635 nm). The DVD is to be compared with the CD that uses an infrared laser (λ = 790 nm). If the storage capacity is inversely proportional to the square of the laser wavelength, this change in laser wavelength gives the DVD a factor of _____ more storage capacity than the CD. If a blue laser (λ = 430 nm) could be used, this factor would increase to _____ .

14. Practice of LPCM digitization on sinusoidal waves with period 8-units vs time. (a) What happens if you digitize the wave every 4 unit? What is the period of the digitized wave? (b) What happens to the digitized wave if you digitized every 6-unit? What is the period of the sampled wave? (c) What happens if the wave is digitized in every 8-unit?

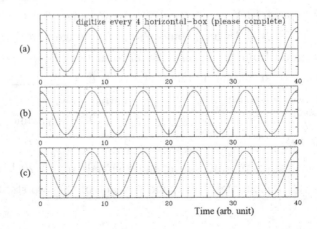

Chapter 9

Room Acoustics

In the free field, the sound pressure and sound intensity decrease with distance from the source as $p \sim 1/r$ and $I \sim 1/r^2$, where r is the distance between the source. The intensity of the sound wave obeys the inverse square law of Eq. (2.23). The sound level decreases 6 dB for each doubling of the distance from the source [see Fig. 9.1(a)]. In a room, a sound wave encounters absorption, reflections, interference, and room resonances. This chapter discusses the basic physic of sound wave propagation in a room. Room acoustic is important in the design of a home, apartment, classroom, theater, concert hall, etc.

I.   Properties of Sound Waves in a Room

In a room, sound waves can reach a listener's ear via reflection by many surfaces of walls. The reflection of sound wave obeys the "law of reflection" [see Eq. (2.9)]. The non-absorbed sound waves still obey the inverse square law after reflection. Figure 9.1(b) schematically shows reflected waves in a room and their sound level vs time. In a typical room with dry walls and concrete or wood floors, sound waves will undergo dozens of reflections before they become inaudible. The sound intensity reaching a listener in a room needs to include wave reflection from the walls. The "direct sound" represents sound directly from source; and the "first reflection" represents the first few reflections from all walls in the room. The sound reaching the listener within 50–80 ms of the direct sound is often called "early sound" [see Fig. 9.1(b)]. After the first group, the reflections become smaller and closer in time together, which is called "reverberate sound."

237

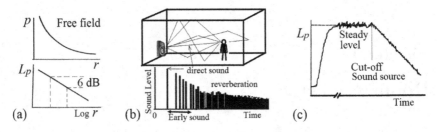

Figure 9.1: (a) The inverse square law of sound wave propagation in a free field. (b) Schematic drawing of a listener in a room, classifying into direct sound, early sound and reverberation. (c) In a room, sound from a steady source will reach an equilibrium sound level. Removing the sound source, the sound intensity will decay exponentially.

For a steady sound source in a room, the sound level can reach an equilibrium when the sound source energy is equal to the room absorption, as schematically shown in Fig. 9.1(c). Removing the sound source, the sound intensity decays exponentially or equivalently the sound level decreases linearly with respect to time.

Consider the uniformly distributed sound energy density of a room in the steady-state condition, shown in Fig. 9.1(c). Let u be the sound wave energy density. The total sound wave energy in the room is $U = uV$, where V is the room volume. The sound intensity is $I = uv$, where v is the speed of the sound. The sound power impinging onto a wall of surface area S is $\frac{1}{4}IS$, where the factor $\frac{1}{4}$ is the sound intensity average over all angles in a hemisphere onto the wall. If there is no sound wave reflection from the wall, the time it takes for sound energy fully absorbed is $\Delta t_{\text{MFP}} = \frac{U}{IS/4} = \frac{4V}{vS}$. This is the average time of a sound wave before a reflection occurs. The mean free path (MFP), defined by Sabine for the average distance between successive reflection, is

$$\text{MFP} = v\Delta t_{\text{MFP}} = 4V/S. \qquad (9.1)$$

★ Example 9.1: Find the mean free path (MFP) of a room measuring $10\,\text{m} \times 10\,\text{m} \times 3\,\text{m}$; and the number of reflections in 1 s.

Answer: The volume is $V = 300$ m^3, and the surface area is $S = 2 \times (10 \times 10 + 10 \times 3 + 10 \times 3) = 320$ m^2; we find MFP $= 4 \times 300/320 = 3.75$ m. If there is no absorption, the time between reflection is $\Delta t_{\text{MFP}} = \text{MFP}/v = 0.0109$ s with $v = 343$ m/s. The number of reflection is $v/\text{MFP} = 91$ reflections per second. Absorption in the room will reduce the number of sound wave reflections.

A. Sabine Area

In a steady state, the acoustic energy distribution in a room is uniform. The air, wall, people and furniture in a room can absorb sound wave. The sound intensity decays exponentially. The decay rate depends on the acoustic attenuation of the room. One defines the absorption coefficient of a surface as

$$a \equiv \frac{I_{\text{absorbed}}}{I_{\text{incident}}}, \qquad (9.2)$$

where $a = 0$ corresponds to 0% absorption or 100% reflection, and $a = 1$ for complete absorption. The sound intensity impinging on the wall surface is $I = uv \cos\theta$, where θ is the incident angle onto the wall with $-\pi/2 \leq \theta \leq \pi/2$. The average flux absorbed by the wall is

$$F_{\text{loss}} = \frac{1}{4} \times uv \times \sum a_i S_i = \frac{1}{4} I A \quad \text{with} \quad A = \sum a_i S_i \qquad (9.3)$$

where u is the sound energy density, v is the speed of sound, $I = uv$ is sound intensity, the factor $\frac{1}{4}$ arises from the average of sound flux onto the wall from all angle θ in the hemisphere, a_i is the absorption coefficient of the surface area S_i, A is called the "Sabine area." The average absorption coefficient of a room is

$$\langle a \rangle \equiv \sum a_i S_i \Big/ \sum S_i . \qquad (9.4)$$

B. Sound Intensity and Sound Level of Sound Sources in a Room

The sound intensity and pressure level at a distance r from a sound source are [see Eq. (2.23)]

$$I_{\text{direct}} = Y \frac{W}{4\pi r^2}, \quad \text{and} \quad L_{p,\text{direct}} = L_{\text{w}} + 10\log \frac{Y}{4\pi r^2}, \qquad (9.5)$$

respectively. Here W is the power of the sound source, L_p is the sound level, L_{w} is the *sound power level* relative to 10^{-12} [W], and Y is the directivity factor. Given a source with sound power W in a room, the power absorbed by the walls is $\langle a \rangle W$ in the first encounter of the wall. The power available for reverberation is $(1 - \langle a \rangle)W$. Equating the reverberation power to the power loss of Eq. (9.3), we find the equilibrium condition: $(1 - \langle a \rangle)W = F_{\text{loss}} = (A/4)\, I_{\text{reverb}}$. Thus, the reverberation sound intensity and its level are

$$I_{\text{reverb}} = \frac{(1 - \langle a \rangle)W}{A/4} \quad \text{and} \quad L_{p,\text{reverb}} = L_{\text{w}} + 10 \log\left(\frac{4(1 - \langle a \rangle)}{A}\right). \quad (9.6)$$

The intensity of reverberation is independent of the location in the room. The combined sound intensity is $I = I_{\text{direct}} + I_{\text{reverb}}$, so the combined sound level is

$$L_p = L_{\text{w}} + 10 \log\left(\frac{Y}{4\pi r^2} + \frac{4(1 - \langle a \rangle)}{A}\right). \quad (9.7)$$

When there are many sound sources in a room, the combined sound level is

$$L_p = L_{\text{w}} + 10 \log\left(\sum_i \frac{F_i Y_i}{4\pi r_i^2} + \frac{4(1 - \langle a \rangle)}{A}\right),$$

where $L_{\text{w}} = 10 \log \frac{\sum W_i}{10^{-12}\text{W}}$, $F_i = \frac{W_i}{\sum W_i}$ is the power fraction of the i-th source, and r_i and Y_i are the distance from and the directivity factor of the i-th source to the measurement position, respectively.

C. Critical Distance

In a reverberant space, the sound perceived by a listener is a combination of direct and reverberant sounds. The "critical distance" is the distance from a sound source that the direct and reverberant sound intensities are equal. Setting $I_{\text{direct}} = I_{\text{reverb}}$ of Eqs. (9.5) and (9.6), we find

$$d_{\text{critical}} = \frac{1}{4}\sqrt{\frac{YA}{\pi(1 - \langle a \rangle)}} \approx 0.057\sqrt{\frac{YV}{(1 - \langle a \rangle)\text{RT}_{60}}}, \quad (9.8)$$

where the unit of d_{critical} is in [meter], RT_{60} in [second], Y is the directivity factor, and the volume V in [m^3]. Note that $d_{\text{critical}} \to \infty$ as $\langle a \rangle \to 1$ in an anechoic room.

★ Example 9.2: Suppose a rectangular auditorium has dimensions $20 \times 30 \times 10$ m and an average absorption $\langle a \rangle = 0.15$. A speaker standing at an edge of a wall radiates 10^{-5} W of acoustical power. What is the SPL at a distance 5 m from the source? What is the critical distance?

Answer: First, a speaker with 10^{-5} W of acoustical power corresponds to $L_{\text{w}} = 70$ dB. The Sabine area of 6 walls is $A = 0.15 \times (20 \times 30 + 20 \times 10 + 30 \times 10) \times 2 = 330$ m^2. Since the speaker is located at a corner, we choose $Y = 4$, and find total sound level is

$$L_p = 70 + 10\log\left(\frac{4}{4\pi \times 5^2} + \frac{4 \times 0.85}{330}\right) = 70 - 16.4 = 53.6 \text{ dB}.$$

Here, the direct sound level $L_{p,\text{direct}} = 70 + 10\log\left(\frac{4}{4\pi 5^2}\right) = 70 - 18 = 52$ dB, and the reverberation of the room is $L_{p,\text{reverb}} = 70 + 10\log\left(\frac{4 \times 0.85}{330}\right) = 70 - 20 = 50$ dB. The total sound level is 53.6 dB. The critical distance is $d_{\text{critical}} = \frac{1}{4}\sqrt{\frac{4 \times 330}{\pi \times 0.85}} = 5.56$ m. At that particular location, both the reverberation and direct levels are 50 dB that give a total sound level 53 dB.

★ Example 9.3: A home listening room has a total Sabine absorption $A = 25$ m^2 with $\langle a \rangle = 0.10$. The source, a loudspeaker placed along one wall, has the directivity factor $Y = 2$ at low frequency and $Y = 10$ at high frequency for a particular direction. What are the critical distances d_{critical} for sound wave at low and high frequencies from the speaker?

Answer: Using Eq. (9.8), the critical distance is $\frac{d_{\text{critical}}}{\sqrt{Y}} = \frac{1}{4}\sqrt{\frac{25}{\pi(1-0.1)}} = 0.74$ m. At low frequency with $Y = 2$, we find $d_{\text{critical}} = 1.0$ m; and at high frequency with $Y = 10$, we have $d_{\text{critical}} = 2.4$ m. This is only an estimation of the critical distance. The Sabine absorption area and the average absorption coefficient may also depend on frequency. Room acoustics is a very complicated subject. A complex model is required to assess the acoustics of a room.

D.　Reverberation Time (RT)

The total acoustic energy in a room is $U = uV$, where u is the acoustic energy density and V is the volume of the room. The average flux absorbed in a room F_{loss} of Eq. (9.3) is proportional to the total energy in the room. Since the acoustic *energy loss rate is proportional to the total energy*, the total sound energy U obeys the exponential decay law:

$$U = U_0 e^{-t/\tau}, \quad \text{where} \quad \tau = \frac{uV}{F_{\text{loss}}} = 4\frac{V}{vA} \qquad (9.9)$$

is the "exponential decay time," $e = 2.71828$ is the natural number of the exponential function (see Appendix B), and U_0 is the initial sound energy at $t = 0$. The reverberation time (RT), defined by W.C. Sabine in 1890 as the time for *a sound energy to decrease its sound level by 60 dB*, is

$$\text{RT}_{60} = \ln(10^6)\tau = 13.8\tau = K\tfrac{V}{A} \qquad (9.10)$$

where $K = \frac{4}{v}(\ln 10^6) = 0.161$ [s/m], $\ln(x)$ is the natural logarithmic function of argument x, $\ln(10^6) = 13.8$, and $v = 343$ m/s for the speed of sound. In imperial units, $K = 0.0490$ [s/ft]. The reverberation time increases with the room size and is proportional to $1/\langle a \rangle$.

E.　Measurement of Reverberation Time

A steady sound source will produce an equilibrium sound energy in a room. Turn the sound source off, and the acoustic energy will decay exponentially. Figure 9.2(a) shows the decay of acoustic energy and the RT_{60}, defined as the time for the intensity level to reduce by 60 dB. If the decay curve does not have the dynamic range of 60 dB, one can measure the time for 30-dB or 20-dB decay to obtain RT_{30} or RT_{20}, and get $\text{RT}_{60} = 2\text{RT}_{30} = 3\text{RT}_{20}$, etc. Figure 9.2(b) shows an example of a "dual" RT_{60} time constant that may signify the effects of acoustically coupled rooms.

Since the absorption coefficient depends on the frequency of the sound, the measured RT_{60} depends on the frequency. In the presence of non-overlapping standing waves (resonances), the decay of acoustic energy may fluctuate with time, as shown schematically in Fig. 9.2(c).

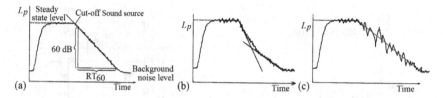

Figure 9.2: (a) Schematic drawing of sound level decay in a room. (b) Schematic drawing of two slopes in the decay curve, associated with acoustically coupled rooms. (c) The decay of acoustic energy may fluctuate with time due to standing wave modes.

Example 4.3 in Chapter 4 lists typical center frequencies of 10-octave-band noise sources for acoustic energy decay measurements. Each octave band can also be refined into $\frac{1}{2}$, $\frac{1}{3}$, $\frac{1}{4}$ or $\frac{1}{6}$ octave bands to achieve a finer resolution of the frequency response of the room. At low frequencies and high resolution, the decay curve of small rooms may exhibit large fluctuation versus time, and the measured RT_{60} may fluctuate versus frequency. The fluctuation of acoustic energy decay depends also on measurement and sound source locations because of standing wave modes.

F. Air Absorption of Acoustic Energy

Air molecules weakly absorb sound energy. The sound intensity I obeys the attenuation law: $I = I_0 \times e^{-\alpha x}$, where x is the distance of propagation, I_0 is the intensity at $x = 0$, α is the attenuation coefficient, usually measured in [dB/100 m]. The attenuation coefficient depends on temperature and humidity, and is generally proportional to f^2, where f is the frequency of the sound wave. Figure 9.3 shows data of air absorption coefficient at various conditions by C.M. Harris. Including the air absorption, the effective Sabine absorption area A_{eff} is

$$\frac{1}{4}A_{\text{eff}} = \frac{1}{4}A + mV \quad \text{or} \quad A_{\text{eff}} = A + 4\,mV, \tag{9.11}$$

where the factor $\frac{1}{4}$ comes from the average of acoustic intensity impinging onto the wall from all directions in a hemisphere. The air absorption constant m has units of $[\text{m}^{-1}]$.

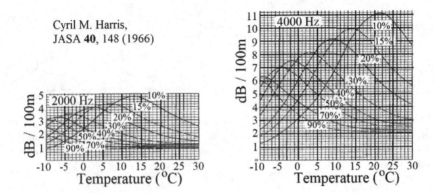

Figure 9.3: Acoustic absorption coefficient of air vs temperature at relative humidity in percent [Data by C.M. Harris, *JASA*, **40**, 148 (1966)].

★ Example 9.4: Use the absorption properties of room materials to calculate RT_{60} for a room $20 \times 15 \times 8$ m³ at $f = 500$ Hz. The walls of the room are painted concrete, the ceiling is plastered on lath, and the floor is carpeted on concrete.

Materials \ Frequency (Hz)	125	250	500	1000	2000
Concrete Block (Dense, Painted)	.10	.05	.06	.07	.09
Carpet (Heavy, on Concrete)	.02	.06	.14	.37	.60
Plaster (Rough Finish on Lath)	.14	.10	.06	.05	.04

Answer: Using the absorption coefficients at 500 Hz, the Sabine areas are respectively: Wall: $A_w = 2(15 \times 8 + 20 \times 8) \times 0.06 = 34$ m²; Ceiling: $A_c = 20 \times 15 \times 0.06 = 18$ m²; Floor: $A_{\text{floor}} = 20 \times 15 \times 0.14 = 42$ m². The total volume of the room is $V = 20 \times 15 \times 8 = 2,400$ m³. The air absorption is negligible at 500 Hz. The reverberation time is $RT_{60} = 0.161V/(A_w + A_c + A_{\text{floor}}) = 4.1$ s.

G. Eyring Absorption Area and Reverberation Time

For an anechoic room, the walls have 100% absorption with $\langle a \rangle = 1$. The reverberation energy should be zero, and the absorption area should be infinity and independent of the room size and geometry. Eyring proposed an alternative definition of RT_{60} as[1]

$$RT_{60} = 0.161 \frac{V}{-S\ln(1 - \langle a \rangle)} = 0.161 \frac{V}{A_{Ey}}, \qquad (9.12)$$

where V is the volume of the room, S is the surface area of the walls and $\langle a \rangle$ is the average absorption coefficient, defined in Eq. (9.4). The absorption area A_{Ey} in the Eyring equation is $A_{Ey} = -S\ln(1 - \langle a \rangle) = \langle a \rangle S(1 + \frac{\langle a \rangle}{2} + \frac{\langle a \rangle^2}{3} + \cdots)$. For small $\langle a \rangle$, $A_{Ey} \approx \sum a_i S_i$, while $A_{Ey} \to \infty$ as $\langle a \rangle \to 1$. Figure 9.4 compares the Eyring effective absorption-coefficient (solid line) with the Sabine average absorption-coefficient (dashed line). One chooses Eyring equation at $\langle a \rangle > 0.25$. Including the acoustic air absorption, the effective absorption area is $A_{\text{eff}} = -S\ln(1 - \langle a \rangle) + 4mV$.

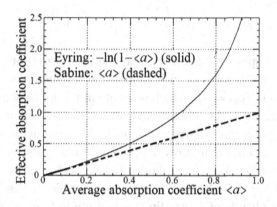

Figure 9.4: Comparison of the Eyring effective absorption coefficient (solid line) with the Sabine effective absorption coefficient (dashed line).

II.   Room Acoustics

We characterize the room acoustics with the Sabine reverberation time RT_{60}. The design criteria of a room or a good concert hall include qualities, such as the RT_{60}, intimacy, liveliness, warmth, clarity, spaciousness, etc. Nevertheless, the RT_{60} is still the single most important parameter for judging the acoustical properties of a room and its suit-

---

[1]Carl F. Eyring, Reverberation time in "dead" rooms, *JASA*, **1**, 217 (1930).

ability for various uses. Following extensive surveys, acousticians and musicians conclude that high reverberation times at 1.8–2.0 s are desirable for music halls, while low reverberation at 0.4–0.6 s is desirable for speech intelligibility in auditoriums. We discuss the effects of standing waves or resonances in a room that may affect the reverberation time.

A.   Normal Modes of a Rectangular Room with Simple Geometry

A small room will have many non-overlapping normal modes in audible range. Consider a room with dimensions L, W, H (length, width and height) for three-dimensional (3D) shoebox geometry. The resonant frequencies for the sound waves in the room are

$$f_{lmn} = \frac{v}{2}\sqrt{\left(\frac{l}{L}\right)^2 + \left(\frac{m}{W}\right)^2 + \left(\frac{n}{H}\right)^2}, \tag{9.13}$$

where v is the speed of sound wave in the room, and integers (l, m, n) are *normal mode* numbers. For example, the resonant frequency for the $(1, 0, 0)$ mode is $f_{100} = \frac{v}{2L}$, which is identical to the fundamental mode of the one-dimensional standing wave of Eq. (2.16). Higher order modes in Eq. (9.13) are space-harmonic modes. Measurements of RT_{60} can identify the existence of non-overlapping normal modes. In the measurement, one should place microphones (detectors) at pressure antinode positions. Figure 9.5 shows a schematic example of the particle displacement amplitude of a two-dimensional mode $(l, m, n) = (2, 2, 0)$.

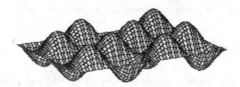

Figure 9.5: Schematic drawing of the molecular displacement amplitude of a 2D normal mode $(l, m, n) = (2, 2, 0)$. Air molecules are not free to move at the walls.

The oscillation amplitude is maximum at the antinodes. The pressure antinodes are located at the displacement nodal positions. The concept of Sabine reverberation time requires uniform sound energy in all direction. The existence of non-overlapping normal modes may violate this condition. A treated room by detuning these sharp resonances will fulfill this condition for achieving a better acoustic quality. Methods of removing these resonances are implementation of sound diffusers, and sound absorbers.

★ Example 9.5: Find the lowest 10 resonant-frequencies of a room with dimension $6 \times 4 \times 2$ m^3. If the average absorption coefficient is $\langle a \rangle = 0.10$, what is the reverberation time?

Answer: The lowest modes are $(l, m, n) = (1, 0, 0)$, $(0, 1, 0)$, $(1, 1, 0)$, $(2, 0, 0)$, $(2, 1, 0)$, $[(3, 0, 0), (0, 2, 0), (0, 0, 1)]$, $[(1, 0, 1), (1, 2, 0)]$, etc. Using Eq. (9.13) with $L = 6$ m, $W = 4$ m, $H = 2$ m, and $v = 343$ m/s at 20°C room temperature, we find that the lowest 10 standing wave frequencies are 19 Hz, 28 Hz, 34 Hz, 38 Hz, 47 Hz, 57 Hz (3 modes), 60 Hz (2 modes). There are 3 modes at 57 Hz and 2 modes at 60 Hz. If the room absorption is $\langle a \rangle = 0.10$, the reverberation time is

$$\text{RT}_{60} = 0.161 \times \frac{6 \times 4 \times 2}{[0.10 \times 2 \times (6 \times 4 + 6 \times 2 + 4 \times 2)]} = 0.69 \text{ s}.$$

B. Mode Number and Mode Density

In a room, the number of modes is

$$N_f = \frac{4\pi}{3} V \left(\frac{f}{v}\right)^3 + \frac{\pi}{4} S \left(\frac{f}{v}\right)^2 + \frac{L'}{8} \frac{f}{v} \approx \frac{4\pi}{3} V \left(\frac{f}{v}\right)^3. \quad (9.14)$$

Here v is the speed of sound, V is the volume, S is the surface area, and $L' = 4(L+W+H)$, or the sum of the lengths of all sides of the room. The first term corresponds to the volume modes. The "spherical volume" in frequency space is $\frac{1}{8} \times \{\frac{4\pi}{3} f^3\}$, where the factor $\frac{1}{8}$ comes from the portion of the sphere with positive frequencies. Since the volume of each standing-wave mode spacing is $\frac{v}{2H} \times \frac{v}{2L} \times \frac{v}{2W} = \frac{v^3}{8V}$, the number of mode becomes $\frac{4\pi}{3} V (\frac{f}{v})^3$, thus the first term in Eq. (9.14) is derived. Using similar arguments, one can derive the second term for the surface modes, and the last term for the linear mode. The number of modes depends mainly on the volume of a room. The number of modes in a frequency bandwidth of Δf is

$$\Delta N_f = 4\pi V \frac{f^2}{v^3} \Delta f + \frac{\pi}{2} S \frac{f}{v^2} \Delta f + \frac{L'}{8v} \Delta f \approx 4\pi V \frac{f^2}{v^3} \Delta f. \quad (9.15)$$

Figure 9.6(a) compares mode densities for rooms having the same volume of $162 \, \mathrm{m}^3$, using room geometries $5.45 \, \mathrm{m} \times 5.45 \, \mathrm{m} \times 5.45 \, \mathrm{m}$; $9 \, \mathrm{m} \times 6 \, \mathrm{m} \times 3 \, \mathrm{m}$; $8.75 \, \mathrm{m} \times 5.25 \, \mathrm{m} \times 3.5 \, \mathrm{m}$; and $7.34 \, \mathrm{m} \times 5.22 \, \mathrm{m} \times 4.24 \, \mathrm{m}$. The speed of sound is assumed to be $v = 343$ m/s. We sample the mode density at the jnd of 3.6 Hz width. These rooms have width, length and height given by $W{:}L{:}H = 1{:}1{:}1$, $3{:}2{:}1$, $5{:}3{:}2$ and the "golden" ratio $(\sqrt{5} - 0.5){:}(\sqrt{5} - 1){:}1$. Their mode densities are essentially the same. The fluctuation seems to be larger for a room at low integer geometric number. However, the mode number fluctuation remains high at low frequency even with golden geometric ratio.

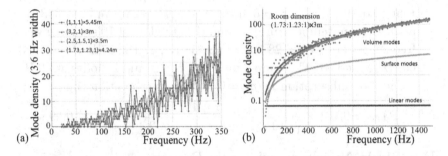

Figure 9.6: (a) The mode density, sampled in 3.6 Hz width, for 4 rooms geometries at the same volume of $162 \, \mathrm{m}^3$. (b) Decomposition of mode density into volume, surface and linear modes in Log-linear plot for one of the room in (a). The numerical simulation data should compare with the sum of all modes.

The mean free path of these rooms are 5.45 m; 7.2 m; 6.56 m; and 6.1 m, respectively. The average times for the sound to bounce wall-to-wall are $t_{\mathrm{MFP}} = 16$ ms, 21 ms, 20 ms, and 18 ms respectively. The sound decay time is $\tau = t_{\mathrm{MFP}}/\langle a \rangle$, where $\langle a \rangle$ is the average absorption of the walls. Figure 9.6(b) compares the numerical results with normal modes decomposed into volume, surface, and linear modes for a room with geometry $(\sqrt{5}\text{-}0.5{:}\sqrt{5}\text{-}1{:}1) \times 3 \, \mathrm{m}$ or $(1.73{:}1.23{:}1) \times 3 \, \mathrm{m}$, at the spectral density up to 1.5 kHz. The total mode density is obtained by summing up the total number of modes within the frequency width of jnd $\approx$ 3.6 Hz. The volume mode density dominates in acoustic resonances. At low frequency, the acoustic resonances are important in the acoustic response function, which also depends on the excitation location of the source, the source frequency contents.

C.   Bandwidth of Acoustic Resonances

Standing wave modes are resonances. The bandwidth Δf of a resonance is the width of the resonance response function at 3 dB below
the maximum. The Q-factor of a resonance is

$$Q \equiv 2\pi \frac{\text{Stored Energy}}{\text{Energy loss in 1 cycle}} = 2\pi f_0 \frac{\text{Stored Energy}}{\text{Energy loss rate}}$$

$$= 2\pi f_0 \frac{\text{Stored Energy}}{\text{Power dissipation}} = \frac{f_0}{\Delta f} .$$

Since the sound energy is $U = U_0 e^{-t/\tau}$ with $\tau =$ (stored energy)/(power
dissipation rate), as shown in Eq. (9.8), the Q-factor is $Q = 2\pi f_0 \tau$ (see
Appendix B(3) for derivation). The Q-factor and the bandwidth of a
resonance become

$$Q = 2\pi f_0 \tau , \quad \text{and} \quad \Delta f = \frac{f_0}{Q} = \frac{1}{2\pi\tau} = \frac{2.2}{\text{RT}_{60}}, \qquad (9.16)$$

where τ is the sound decay time given by Eq. (9.8). The resonance bandwidth shown in Eq. (9.16) increases with the average absorption coefficient. A sharper resonance will have a larger Q-factor and a longer reverberation time. The existence of isolated standing wave modes produces
large fluctuation on RT_{60}. The intensity is proportional to square of the
wave amplitude; thus, the wave amplitude time-constant is $\tau_A = 2\tau$, we
find $Q = \pi f_0 \tau_A$. The resonance bandwidth becomes $\Delta f = \frac{1}{\pi \tau_A}$.

★ Example 9.6: What is the resonance width of the room resonance
   discussed in Example 9.5 if the average absorption is $\langle a \rangle = 0.10$?

   Answer: The room in Example 9.5 has $\text{RT}_{60} = 0.69$ s. With
   Eq. (9.16), the resonance width of the room is $\Delta f = \frac{1}{2\pi\tau} = \frac{2.2}{\text{RT}_{60}} = \frac{2.2}{0.69} = 3.2$ Hz, which is smaller than the separation of low-
   frequency modes (see Example 9.5 for frequencies of 10 low order
   modes). The room acoustics receives enhancement at frequencies of
   the non-overlapping room resonances. If you speak in a small room
   without any furniture, you can experience resonant sound effects
   with enhancement and echo of sound waves at certain low frequencies. The bandwidth is larger at higher absorption $\langle a \rangle$.

D. Frequency Response of a Small Room and the Schroeder Frequency

In the study of statistical distribution of normal modes on room acoustics, M.R. Schroeder divided the frequency space into four regions according to their response of resonance characteristics.[2]

Region I: Low-lying harmonics dominate the acoustics of a small room in the low-frequency region. It is particularly important if the fundamental standing wave mode frequency

$$f_1 = \frac{v}{2L} \tag{9.17}$$

is above our hearing range of above 20 Hz, where v is the speed of sound and L is the longest length of a room. Any frequency below f_1 does not excite resonances in the room. Since the low-lying room resonances are non-overlapping (see Example 9.6), room resonances with frequency higher than f_1 can cause sound timbre distortion. This is most relevant to the acoustics of small rooms with small absorption $\langle a \rangle$.

Figure 9.7 shows the lowest fundamental mode frequencies f_1 (dashed and dash-dotted lines) vs volume for a room with a fixed ceiling height at 3 m, while changing the width and length. Since the room height is 3 m in this example, the vertical mode frequency is constant. If the frequency dependence of the average absorption coefficient is small, the RT$_{60}$ does not depend on frequency. Assuming the average absorption $\langle a \rangle = 0.15$, we find the resonance bandwidth to be about 3 Hz. Most low-frequency modes are non-overlapping for small rooms.

Region II: As the frequency increase, the density of resonance also increases. When many resonances overlap, the resonance structure is washed out, and the spectral distribution becomes Gaussian. One can describe the acoustic wave with statistical properties. The

---

[2]M.R. Schroeder, Die statistischen Parameter der Fre-quenzkurven von grossen Räumen, *Acustica*, **4**, 594 (1954).

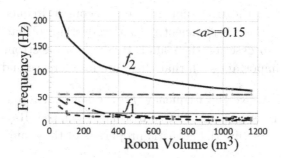

Figure 9.7: The f_1 and f_2 vs the room size based on the geometry of rooms with a fixed height of 3 m with varying length and width. The 3 fundamental modes f_1 (length, width, height; dashed, dash-dotted lines) are shown for comparison with the 20 Hz thin solid line.

"crossover frequency" discovered by M.R. Schroeder is

$$f_2 \sim 2{,}000\sqrt{\frac{\mathrm{RT}_{60}}{V}} \approx \frac{800}{\sqrt{A}}, \text{ for } (V \text{ and } A \text{ are in m}^3 \text{ and m}^2). \quad (9.18\mathrm{a})$$

$$f_2 \sim 11{,}885\sqrt{\frac{\mathrm{RT}_{60}}{V}} \approx \frac{2624}{\sqrt{A}}, \text{ for } (V \text{ and } A \text{ are in ft}^3 \text{ and ft}^2). \quad (9.18\mathrm{b})$$

The crossover frequency is the critical frequency that statistical properties apply. Using the bandwidth Δf of Eq. (9.16), we find that the number of modes is $\Delta N_f \approx 4\pi V \frac{f^2}{v^3}\frac{1}{2\pi\tau} = \frac{1.11\times10^8}{(343)^3} \approx 2.7$. There are about three overlapping resonances within each resonance bandwidth, *independent of the absorption properties of the room* and independent of room size. Figure 9.7 shows the crossover frequency decreasing with the average absorption coefficient vs room size. For a typical small room, f_2 lies between 100 and 200 Hz. Both f_1 and the Schroeder crossover frequency f_2 decreases as the room size increases.

In the frequency range between (f_1, f_2), an individual resonance may dominate room acoustics. The response may depend on the location, frequency content of the source and measurement location. Modifying the room acoustics by using diffusers and sound absorbers is important in the design of audio-recording rooms.[3]

---

[3] A memorial volume in honor of Manfred R. Schroeder is available at https://link. springer.com/content/pdf/10.1007%2F978-3-319-05660-9.pdf

Region III: In the transition region at acoustic frequency range of $(f_2, f_3 \approx 4f_2)$, there is sufficient resonance overlap so that one can use ray-acoustics to describe the sound. However, some strong resonances may still be important such that there are interference and diffraction.

Region IV: Above the frequency f_3, the spectral resonances highly overlap. There is no dominant individual resonance. Ray-acoustics is applicable to describe the acoustic properties of the room.

E. Concert Halls

Big concert halls normally have irregular geometry. Their acoustic properties require detailed computer modeling and measurements. Based on a survey of acoustic researchers and musicians, desirable reverberation times (in seconds) for various functions vs the auditorium size are shown in Fig. 9.8(a). For speech clearness, auditoriums are designed to have short reverberation time.

Musicians and audience seem to prefer an optimal reverberation time of about $T_m \approx 1.6$ s for a music concert hall. Furthermore, the preferable reverberation time should increase at low frequency to enhance the base for a good concert hall, as shown in Fig. 9.8(b). Recent work has stressed the importance of having sufficient reflected sound arriving from the sides. Such lateral reflections with time delay from 25–80 ms add the feeling of spaciousness, whereas overhead reflections during the same time delay add mainly to the early sound. The spatial impression requires a sufficient portion of the early sound to arrive from the sidewall

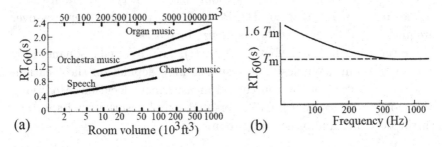

Figure 9.8: (a) Preferred RT_{60} for music and speech clearness based on a survey of professional musicians. (b) Preferred RT_{60} vs frequency for good concert halls.

reflections, in order to increase the auditory source width so that the reverberant sound appears to come from all directions.

In general, good acoustics for concert halls, opera houses, lecture halls, theaters, and churches are quite different. Common criteria for good acoustics are *adequate loudness, uniformity, clarity, reverberance or liveliness, freedom from echoes, and low level of background noises.* The topic is beyond this introductory text on the basic physics of sound. An extensive treatment of room acoustics is available in a book by F. Alton Everest and Ken Pohlmann (see Ref. [12]).

III.   Summary

This chapter discusses sound propagation and reverberation in a room. In the presence of a steady sound source, the sound energy in the room will reach an equilibrium value. Once the sound source is removed, the sound energy in the room decays exponentially. The decay time $\tau = 4v/vA$ depends on the room volume V, Sabine absorption area A, and the speed of sound in the room v. The Sabine reverberation time $RT_{60} = 13.8\tau$ is the time it takes for sound energy to reduce 60 dB from equilibrium. RT_{60} increases with room size and decreases with the average sound-absorption coefficient. It is the most important number in characterizing the room acoustics.

In a room, a sound wave can interfere to produce standing wave (resonance) modes. The number of standing wave modes is proportional to the room volume and cubic power of frequency. The quality factor of a resonance is $Q = 2\pi f_0 \tau$, where f_0 is the mode frequency and τ is the acoustic energy decay time. The bandwidth of the mode is $\Delta f = f_0/Q = 2.2/RT_{60}$. Low-order room resonances are non-overlapping, and these resonances can enhance reverberant sound and cause frequency distortion in the room. The crossover Schroeder frequency, $f_2 \sim 2000\sqrt{RT_{60}/V}$, decreases with both the average absorption coefficient and the room size. The number of modes within the bandwidth at the crossover frequency is $\Delta N_f \approx 2.7$, independent of the room size and absorption coefficient. Room acoustics is an important, special topic relevant to music hall and auditorium. Realistic simulation tools for room acoustics are available on market and are part of the music hall design (see Ref. [12] for details).

IV.   Homework 09

1. Consider an office measuring 4 m × 4 m × 2 m. Provide answer to the following questions.

   a) Find the mean free path (MFP) and find the number of sound reflections in 1 s.

   b) Find the Sabine area, neglecting acoustic air absorption and assuming no furniture with average absorption coefficient $a = 0.10$ for the walls, floor and ceiling.

   c) Calculate the RT_{60} using Sabine and Eyring formula [see Eqs. (9.8) and (9.11)].

   d) Calculate the critical distance when a speaker is fixed on a wall.

   e) Find the frequencies of the lowest 3 modes at the 20°C.

   f) What is the frequency width Δf of these acoustic resonances of the room?

   g) Find the Schroeder frequencies f_1, f_2, and f_3 of the room.

2. Consider a classroom measuring 10 m × 10 m × 3 m. Provide answer to the following questions.

   a) Find the mean free path (MFP) and find the number of sound reflections in 1 s.

   b) Find the Sabine area, neglecting acoustic air absorption with an effective absorption coefficient $a = 0.30$ for the walls, floor and ceiling, when fill with students in the room.

   c) Calculate the RT_{60} using Sabine and Eyring formula [see Eqs. (9.8) and (9.11)].

   d) Calculate the critical distance when a speaker is fixed on a wall.

   e) Find the frequencies of the lowest 3 modes at the 20°C.

   f) What is the frequency width Δf of these acoustic resonances of the room?

   g) Find the Schroeder frequencies f_1, f_2, and f_3 of the room.

Chapter 10

Noise, Effects on Human Hearing and Regulations

Sound wave is indispensable to human cultures in art and communication, but it can also be annoying. Noise is unwanted sound. Sound sources includes vibrating solid bodies, vibrating air columns, fluid flow, disturbance to the pressure in media, rapid change of temperature or pressure, and objects moving at supersonic speed. All loud sound waves, including pleasant music sound, can induce damage to our hearing system. This chapter discusses sound sources and sound propagation, effects of excessive noise, hearing loss, hearing protection regulations, and noise control and reduction.

I. Environmental Noise Measurements and Evaluation

A. Noise Level Weightings and Measurements

The unit for sound loudness level is the decibel. The loudness level at different frequency may have different physiological effects on our hearing system. The regulatory commission introduced frequency-weighted filter functions A, B, or C for assessing human ears. These filter functions resemble the ISO-226 equal loudness level curve of 40, 70 and 100 phon respectively. Table 10.1 and Fig. 10.1 show some weighting functions, e.g. A 20 dB sound level at 100 Hz is 0 dBA, etc. The D-weighted function emphasizes the frequencies around 1–3 kHz for aerodynamics applications. The B-weighted function is not commonly used. A recent ISO recommendation is the use of A-weighting for commercial aircraft noise. The notation for A-weighted sound pressure level is

Table 10.1: Center frequencies of 10 bands used in survey, measurement, and data analysis.

Weighted	Frequency f_c (Hz)									
Level (dB)	31.5	63	125	250	500	1,000	2,000	4,000	8,000	16,000
dB(A)	-39.4	-26.2	-16.1	-8.6	-3.2	0	$+1.2$	$+1.0$	-1.1	-6.6
dB(C)	-3.0	-0.8	-0.2	0	0	0	-0.2	-0.8	-3.0	-8.5

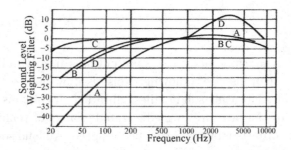

Figure 10.1: The weighting of the dBA/B/C/D scales. For example, a sound level at $L_p = 60$ dB at 100 Hz is equivalent to $60 - 20 = 40$ dBA, $60 - 5 = 55$ dBB, or 60 dBC.

"$L_p(A)$." The Flat-weighting, "L(Linear)-weighting" is now superseded by the "Z(zero)-weighting" for no-frequency weighting.

A sound level meter can carry out frequency analysis and measure the sound level at various frequency bands. The frequency content of noise is normally measured using octave band and $\frac{1}{3}$-octave-band filters; frequency detail is obtained by FFT analysis. Table 10.1 lists the center frequencies of the standard octave bands. The cutoff frequencies of the i-th octave band at center frequency f_{ci} are $2^{-1/2}f_{ci}$ and $2^{1/2}f_{ci}$, respectively. At the cutoff frequency, the filter response is 3 dB below the maximum. These 10 octave bands cover the entire audible frequency range of human hearing. To reach a finer frequency resolution in measurements, we can use $\frac{1}{3}, \frac{1}{6}, \frac{1}{12}$ or $\frac{1}{24}$ octave bands that divide each octave band into 3, 6, 12 or 24 bands.

One can combine the sound levels of each band to obtain a proper sound level of the combined band. Given a sound level L_i of band i, the rms pressure is $p_i^2 = p_0^2 10^{L_i/10}$ [see Eq. (3.5)], where the reference rms pressure is $p_0 = 2 \times 10^{-5}$ [Pa]. The total intensity of the sound is the sum of the intensities of all bands. The combined rms sound pressure is $p_T^2 = \sum_i p_i^2 = p_0^2 \sum_i 10^{L_i/10}$, and total sound level of all bands becomes

$$L_T = 10 \log \frac{p_T^2}{p_0^2} = 10 \log \left(\sum_i 10^{L_i/10} \right)$$

$$= 10 \log \left(10^{L_1/10} + 10^{L_2/10} + \cdots \right). \qquad (10.1)$$

★ Example 10.1: Divide an octave band, centered at 500 Hz, into three $\frac{1}{3}$-octave bands. Find the frequency range of the $\frac{1}{3}$-octave bands.

Answer: When divide the octave band at f_0 into three $\frac{1}{3}$-octave bands; the center frequencies are located at $2^{-\frac{1}{3}} f_0$, f_0, and $2^{\frac{1}{3}} f_0$. Setting $f_0 = 500$ Hz, we find center frequencies at $400\,(397)$, 500, and 630 Hz. The bandwidths of these $\frac{1}{3}$-octave bands are $(2^{-\frac{1}{6}} \times 400, 2^{\frac{1}{6}} \times 400)$, $(2^{-\frac{1}{6}} \times 500, 2^{\frac{1}{6}} \times 500)$, and $(2^{-\frac{1}{6}} \times 630, 2^{\frac{1}{6}} \times 630)$, respectively. The combination of the three $\frac{1}{3}$-octave bands is the 500 Hz octave band.

★ Example 10.2: The sound levels of the $\frac{1}{3}$ octave band at 400, 500, and 630 Hz are 72, 74 and 76 dB. What is the sound level of the octave band at 500 Hz?

Answer: As above, the center frequencies of the three $\frac{1}{3}$-octave bands around 500 Hz are 400, 500 and 630 Hz. Using Eq. (10.1) to combine data of the three $\frac{1}{3}$-octave bands, we find the combined level of the 500-Hz octave band to be $L_T = 10 \log \left(10^{7.2} + 10^{7.4} + 10^{7.6} \right) = 79$ dB.

★ Example 10.3: The Table below lists data of the sound level for a noise source in using the octave bands. What is the overall sound level of the source?

Center Frequency (Hz)	125	250	500	1,000	2,000	4,000	8,000
Sound level (dB)	79	80	94	100	94	94	88

Answer: Since the bandwidths in the table overlap, we can use Eq. (10.1) to find the overall sound level as $10 \log \left(10^{7.9} + 10^{8.0} + 10^{9.4} + 10^{10} + 10^{9.4} + 10^{9.4} + 10^{8.8} \right) = 103$ dB.

B.   Energy Equivalent Level and Day-Night Equivalent Level

There is evidence that damage to our ears is related to the total acoustic energy dosage deposition. This is the equal energy principle. The energy equivalent level, L_{eq}, is the sound level with the same total energy of a steady-state noise over a given period of time, i.e.

$$L_{eq} = 10 \log \left(\frac{1}{T} \sum \frac{p_i^2}{p_0^2} \Delta t_i \right) = 10 \log \left(\frac{1}{T} \sum_i 10^{L_i/10} \Delta t_i \right)$$

$$= 10 \log \left(\frac{1}{T} \sum_i I_i \Delta t_i \right) , \tag{10.2}$$

where $T = \sum_i \Delta t_i$ is the total exposure time. This is the basis of the 3 dB exchange rate in the NIOSH recommendation: an increment of sound level by 3 dB and a reduction in time duration by $\frac{1}{2}$ will achieve the same total energy deposition. Their effect on our hearing system will be similar (see Sec. 10.V).

Noise at nighttime is more annoying. The day-night equivalent level is to add 10 dB to the energy equivalent level, i.e. $L_{dn} = L_{eq} + 10$ dB from 10 p.m. to 7 a.m. in the averaging process.

★ Example 10.4: Use the energy equivalent level to justify the statement: "staying at 120 dB environment for 3.6 s is equivalent to staying at 100 dB environment for 360 s."

Answer: The intensity ratio between 120 dB and 100 dB environments is $I_2/I_1 = 10^{(120-100)/10} = 100$. The energy equivalent time-duration in 100 dB room is $(3.6\,\mathrm{s}) \times (100) = 360$ s.

★ Example 10.5: What is a day-night equivalent level of steady noise at 50-dB noise? (Assume that nighttime has a total of 9 hours, while daytime, including evening, is 15 hours.)

Answer: Using Eq. (10.2), we find $L_{dn} = 10 \log \frac{1}{24}(15 \times 10^{50/10} + 9 \times 10^{(50+10)/10}) = 56$ dB.

★ Example 10.6: You attend a rock concert at 100 dBA for 2 hours and stay in a quite environment for 8 hours at 50 dBA. What is the average intensity in 10-hour period?

Answer: The average intensity in 10-hour period is $\frac{I_{av}}{I_0} = \frac{1}{10}\left(2 \times 10^{\frac{100}{10}} + 8 \times 10^{\frac{50}{10}}\right) = 2.0 \times 10^9$. The average SIL in the 10-hour period is $L_{av} = 10 \log \frac{I_{av}}{I_0} = 10 \log(2.0 \times 10^9) = 93$ dBA. Modern sound level devices can measure the time-averaged sound intensity.

C. Sound Level at Percentage of Time: $L_{x\%}$, Loudness at $x\%$ of Time

The average sound level may not be a useful representation of noise effect. Since the sound level at a given location may uncontrollably depend on time, one way to describe noise is expressed in $L_{x\%}$, or the level exceeded $x\%$ of the time. For example, the sound pressure level $L_{10\%}$ is the level exceeded 10% of the time, and $L_{50\%}$ exceeded 50% of the time. In environmental protection, the noise level is evaluated using the values $L_{10\%}$, $L_{50\%}$, and $L_{90\%}$ the A-weighted sound pressure level exceeded 10%, 50% and 90% of the time. Figure 10.2 shows a schematic measurement of the A-weighted SPL vs the time for a total time period T, e.g. 24 h. The measured data can be used to obtain the percentage of time above a certain SPL and deduce the $L_{10\%}$, $L_{50\%}$, and $L_{90\%}$. This measurement is useful for applications in residential areas near high-traffic highway or near marketplaces.

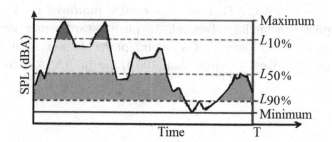

Figure 10.2: A schematic drawing of A-weighted SPL vs time for a period T, e.g. 24 h.

II. Noise Sources

Today, most noise sources come from modern mechanical tools and machines, (e.g. cooling fans), fluid flow, mechanical friction, turbulence, motor vehicles, etc. We list some examples below.

A. Flow Noise

We can categorize periodic fluid-flow noise source into monopole, dipole, quadrupole or higher multipole sources. Monopole sources resemble an expanding and contracting sphere so that the pressure wave is expanding outward uniformly. The dipole noise source arises from two out-of-phase monopole sources. Most noise sources are complex multipole sources. The sound power is proportional to u^4, u^6, and u^8 for the monopole, dipole, and quadrupole sources, respectively, where u is the *flow velocity*. Examples of flow noises are siren, cooling fans, moving stream encountering an obstruction (particularly a sharp object for the edge tone). The best way to minimize noise is high volume, low-velocity flow.

B. Machine Noise

Machine noise arises mostly from mechanical vibration and friction. Methods for reducing the sources of noise include balance of the rotating parts; maintenance of bearings and gears; detuning resonances from noise radiating panels; covering the noise sources by mufflers; etc. Figure 10.3(a) shows the mechanical power vs the A-weighed sound power surveyed by E.A.G. Shaw.[1] Normally, improved efficiency provides a quieter technology. Technologies push toward higher power with less acoustic power output. The sound power output from airplane engine has improved by more than 100 times in NASA quite engine program.

---

[1] Edgar A.G. Shaw, Noise pollution—what can be done? *Physics Today*, **28**(1), 46 (1975); doi:10.1063/1.3068772

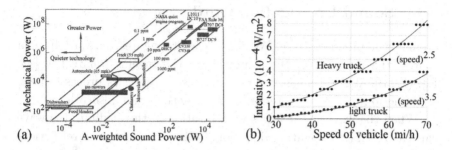

Figure 10.3: (a) Mechanical power vs A-weighted sound power, summarized by A.G. Shaw. (b) Data of measured noise intensity from a truck vs speed, measured at 50 feet from the roadway. Lines are simple curve fitting to the measured data with power law in speed.

C.   Motor Vehicle

In 2016, a DOT report established more than 268.8 million registered vehicles in the US. Automobiles generate Mega-Watt of acoustic power, particularly in cities and highway areas. The noise from automobiles engine/vehicle type, the muffler, traffic operation (e.g. speed), road surface type, tire-type (tires were completely re-designed in 1970 to reduce noise), roadway geometry, micrometeorology, area geometry and structure, and residential setback, etc. Figure 10.3(b) shows the noise intensity of truck vs speed, measured at 50 feet from the roadway, by J.P. Cowan in Ref. [35]. The noise intensity is proportional to a high power in vehicle speed.

Highway noise is an important issue in our society. The noise level of motor vehicles on communities depends on sound barrier, diffraction, reflection, and refraction.[2] Acoustic models were successfully used in the widening project of the NJ turnpike from 6 to 12 lanes and Interstate 66 through Arlington, Virginia.

---

[2] See http://www.fhwa.dot.gov/environment/noise/measurement/mhrn00.cfm

III.   Sound Propagation Outdoors

Sound is a pressure wave. As the wave propagates outward from a source, the intensity of the sound obeys the inverse square law [see Eq. (2.23)]. Wave propagation depends also on physics processes such as refraction, reflection, interference, diffraction, and absorption.

When the wave speed varies with position in the medium, the wave can change its propagation direction. This phenomenon is refraction. The speed of sound depends on air temperature in atmosphere. Normally, the temperature decreases vs altitude, and sound will be refracted upward. Temperature inversion is the phenomenon wherein temperature close to the ground is colder and temperature higher up is warmer. When temperature inversion happens, the speed of sound at high altitude is faster than its speed close to the ground, and the sound wave bends back towards the ground. In the early morning, the temperature inversion occurs at the low altitude, and the sound at small angle partially refracts downward. During a clear night, the temperature inversion occurs at a higher altitude; most sound then refracts downward except sound towards large angles.

Interference can also be important for wave propagation outdoors. Figure 10.4(b) shows a schematic plot of relative sound level vs frequency. Constructive or destructive interference depends on the path difference between the direct path D and the reflected path R, e.g. $R - D = \frac{1}{2}\lambda, \frac{3}{2}\lambda, \frac{5}{2}\lambda, \ldots$ for destructive interference. The actual atten-

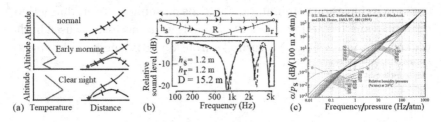

Figure 10.4: (a) Schematic drawing of sound waves propagation due to refraction. (b) Interference can be constructive and destructive for sound waves from the path difference between the direct path D and the reflected path R. The bottom plot shows schematic drawing of excessive attenuation vs frequency. (c) Attenuation of sound wave vs air temperature for various frequencies, and relative humidity.

uation depends on frequency, absorption coefficients, impedance of the media, etc.[3]

Diffraction is the phenomenon that causes the sound wave to bend and spread around the obstacles. The diffraction of the sound wave becomes important when the wavelength of the sound wave is of the same order as that of the obstacle.

Air molecules weakly absorb sound energy, obeying the intensity attenuation law: $I = I_0 \times e^{-\alpha x}$, where α is the attenuation coefficient, usually measured in [dB/(100 m)]. The attenuation depends on frequency f^2. The attenuation coefficient depends also on temperature and humidity. After many experiments and theoretical studies, *accurate* calculation of attenuation coefficient of a sound wave in air is now available.[4] Figure 10.4(c) shows the attenuation coefficient (α/p_s) vs frequency for various humidity, where p_s is the atmospheric pressure.[5] The attenuation of sound wave in seawater is also of great importance to sonar communication.

IV.   Effect of Excessive Sound Level

Hearing loss is bad for health. A major cause of hearing loss arises from exposure to excessive sound levels that can cause conductive and sensorineural hearing loss. Other causes of conductive hearing loss are earwax build up, abnormal bone growth in the middle ear, middle ear infections, and mechanical traumas (e.g., eardrum rupture and dislodgement of the middle ear bones resulting from instantaneous high-level noise like gunshots, explosions, or extreme shockwaves). Sensorineural hearing loss can be congenital, aging, repeated exposure to excessive sound level, acoustic trauma, disease and infection, medication, and tumors. A very small percent of hearing impairment in human new-

---

[3]T.F. Embleton, *JASA*, **100**, 31 (1996); J.E. Piercy, T. Embleton and L. Sutherland, *JASA*, **61**, 1403 (1977); T. Embleton, J.E. Piercy and N. Olson, *JASA*, **59**, 367 (1976).

[4]See e.g. Harris, *JASA*, **40**, 148 (1966); and Fig. 9.3 in Chapter 9.

[5]H.E. Bass *et al.*, *JASA*, **97**, 680 (1995); ANSI Standard S1-26:1995, or ISO 9613-1:1996.

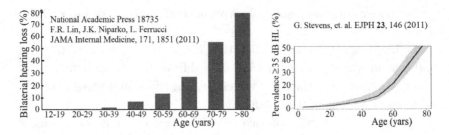

Figure 10.5: Survey of the prevalence of hearing loss in US population (See the review in the National Academic Press 18735 and 23446 at https://www.nap.edu/), and a survey of the prevalence of hearing loss of the world population, carried for the WHO.

borns come from rare genetic mutations. Figure 10.5 shows the survey of hearing loss prevalence of population in the US and in the world.[6]

A. Physiological and Psychological Effects of Noise Exposure and Sleep Disturbance

Noise can cause physiological and/or psychological problems. Exposure to extremely high noise levels can also cause headaches, irritability, fatigue, constricted arteries, and a weakened immune system.[7] The physiological effects includes a wide variety of symptoms, e.g. (1) cardiovascular effects, such as increased heart rate and blood pressure, (2) higher breathing rate, (3) annoyance and aggression, (4) hypertension, (5) high stress levels, (6) tinnitus and hearing loss, (7) sleep disturbances, and (8) muscle contractions and perspiration. Stress and hypertension are leading causes of health problems. Hearing loss can cause depression, dementia and even heart disease. A single 8-hour period exposure to moderately high sound intensity can cause a statistical rise in blood pressure about 5–10 mm-Hg due to vasoconstriction. Psychological effects of hearing loss include nervousness, tension, anger

---

[6]https://www.nidcd.nih.gov/health/statistics/quick-statistics-hearing; F.R. Lin, J.K. Niparko, L. Ferrucci, Hearing loss prevalence in the United States, *JAMA*, *Arch Intern Med.*, **171**, 1851 (2011); and G. Stevens, S. Flaxman, E. Brunskill, M. Mascarenhas, C.D. Mathers, and M. Finucane, *EJPH*, **23**, 146 (2011), https://doi.org/10.1093/eurpub/ckr176.

[7]See http://www.nidcd.nih.gov/health/hearing/pages/noise.aspx.

and irritability. Up to 90% people with tinnitus (a ringing, buzzing, or roaring in the ears or head) have some level of noise-induced hearing loss (NIHL).

Noise can cause sleep disturbance. Noise exposure (including daytime exposure) can cause sleep disturbance at night. Some studies report that: (1) Noise levels of 60 dB wakes 90% of people after they have fallen asleep. (2) Noise levels of 55 dB affects REM cycles and increases time to fall asleep. (3) Noise of 40–45 dB wakes 10% of people. WHO recommends an ambient noise levels below 35 dB for optimum sleeping condition (see Ref. [40]).

B. Temporary Threshold Shift and Temporary Hearing Loss

Frequent exposure to noise may cause hearing loss that can be temporary or permanent. Long time exposure of loud noise also causes temporary threshold shift (TTS). Temporary hearing loss may take 16 to 48 hours to recover. Although the temporary hearing loss seems to disappear, long-term damage to our hearing system persists. We might ignore the warning signs of hearing loss because the damage from noise exposure could be gradual and minute. Figure 10.6(a) shows the *hypothetical* growth in threshold shift after exposure to noise of varying level and duration (left) and the recovery temporal threshold shift (right).[8] Exposure of a noise level 90 dB for 24 hours produces 51 dB TTS. Removing the noise source, it will take 7 days to recover with 11 dB permanent threshold shift (PTS), i.e. noise induced permanent threshold shift (NIPTS). Long time exposure of noise level higher than 90 dB will produce PTS. Unlike bird and amphibian, human hair cells do not grow back. Damage and eventual death of hearing hair cells is the major source of noise-induced hearing loss (NIHL).

Figure 10.6(b) shows the survey data of TTS and its recovery decay in 24 hours at 75–88 dB sound level.[9] Long time exposure of noise will induce an asymptotic threshold shift (ATS). The growth time of ATS depends on the noise level. However, it is difficult to predict PTS from

---

[8] See J.D. Miller, Effects of noise on people, *JASA*, **56**, 729 (1974).

[9] John H. Mills, Robert M. Gilbert, and Warren Y. Adkins, *JASA*, **65**, 1238 (1979).

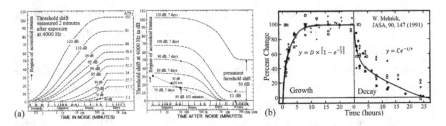

Figure 10.6: (a) Hypothetical illustration of threshold shift with noise exposure of 4000 Hz. (b) Survey data show the TTS and exponential function recovery feature at 75–88 dBA level.

TTS.[10] It is illegal to carry out psychoacoustic experiments on people with sound level 90 dB more than 8-hr duration. It is impossible to verify the hypothetical TTS and PTS curve on Fig. 10.6(a). Neither growth nor recovery of TTS appears to receive benefits from drugs, medication, time of days, hypnosis, or the state of mind. Low-frequency noise produces less TTS.

C.   Permanent Hearing Loss

Two main types of permanent hearing loss are conductive hearing loss and sensorineural hearing loss. The rate of the congenital hearing loss in newborn is about 0.17%. Conductive hearing loss comes from defects in the outer or middle ear, preventing sound from passing efficiently to the cochlea in the inner ear. Examples are buildup of earwax, rupture of eardrum due to a sudden loud sound, otosclerosis, etc. Medical surgery may restore some conductive hearing loss. Sensorineural hearing loss arises from damage to the receptor organs in the inner ear. Figure 10.7(a) shows varying degree of damage to hair cells. Possible causes of hair cell damage come from prolonged exposure to loud sounds, aging, illness, head trauma, toxic medications, or genetic defects.

The *audiogram* is a hearing test that plots the softest sound level a person can hear vs frequency, normally at $\frac{1}{3}$-octave bands from 20 Hz to 16,000 Hz, based on ISO 389-7:2005. Varying degree of hearing loss

---

[10]See W. Melnik, *JASA*, **90**, 147 (1991).

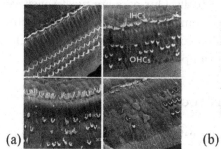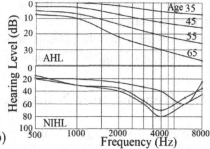

(a)   (b)

Figure 10.7: (a) Normal and varying degree of degradation of hair cells. Loss of hair cells can cause hearing loss. (b) Audiograms of AHL (top) and varying degree of NIHL (bottom).

will produce different shapes of audiogram.[11] Congenital hearing loss will show high hearing level for all frequencies. Beside the congenital hearing loss, Fig. 10.7(b) shows examples of the audiogram for age-related hearing loss (top) and for noise induced hearing loss, which has a dip between 4000 and 8000 Hz (bottom).

(1) Presbycusis or Age-related Hearing Loss (AHL)

Hearing loss is a natural course of aging. Presbyacusis or presbycusis, coined by Zwaardemaker in 1891, is the natural hearing loss due to aging. Most people do not hear much above 8000 Hz at age 70. The top plot of Fig. 10.7(b) shows a typical audiogram for population at different ages. Presbycusis is irreversible and incurable. Hearing aids can provide relief for the needs of daily life.

(2) Noise Induced Hearing Loss (NIHL)

Exposure to loud sounds can also cause hearing loss, and the damage is cumulative. People working in noisy environment can suffer severe hearing loss called noise induced hearing loss (NIHL). The audiogram

---

[11]Schuknecht H.F., Gacek M.R., Cochlear pathology in presbycusis, *Ann. Otol. Rhinol. Laryngol.*, **102**, 1 (1993); George A. Gates, John H. Mills, *Lancet*, **366**, 1111 (2005); Kelly Demeester *et al.*, *International Journal of Audiology*, **48**, 222 (2009); etc. for survey data and analysis.

of NIHL typically shows a dip at 4000 Hz. Our ears are most sensitive to the frequency near 4000 Hz and the muscle fatigue of larger vibration at the cochlear membrane causes damage to the hearing hair cells. The bottom plot of Fig. 10.7(b) shows examples of NIHL.

V. Hearing and Noise Protection Regulation

State and provincial governments have established various form of noise guideline throughout history. Japan passed somewhat limited noise regulation in 1968. The United States passed a more comprehensive Noise Control Act (NCA) in 1972 and the Quiet Communities Act (QCA) in 1978. The acts restrict noise level, noise duration, noise sources, and the time of noise. The law established a national policy to promote an environment free from noise that may harm health and welfare. The Acts aimed to coordinate federal research and activities in noise control; establish noise emission standards; and inform the public on noise emission and noise reduction characteristics. In 1981, the federal government unfunded the office and transferred the responsibility to local and state governments, but Congress never repeals the NCA and QCA, and they remain in effect. World governments and WHO have established similar noise control acts.

A. OSHA Noise Emission Regulation (Government Law)

The Occupational Safety and Health Administration (OSHA) must consider technical and economic feasibility and set permissible limit of daily dose for workers. OSHA set the Permissible Exposure Level (PEL) at 90 dB 8-hour time-weighted average (TWA), with 5 dB exchange rate, shown in Table 10.2. The permissible time T_{PEL} for A-weighted average loudness level L is

$$T_{\text{PEL}}(L) = 8 \times 2^{(90-L)/5} \ [\text{hours}]. \qquad (10.3)$$

The effect of exposure at different environments with different sound level is

$$\text{Fraction of dose} = \frac{t_1}{T_1} + \frac{t_2}{T_2} + \cdots + \frac{t_n}{T_n}. \qquad (10.4)$$

Table 10.2: The 5-dB exchange rate table set the OSHA 1910.95. Details are available at https://www.osha.gov/pls/oshaweb/owadisp.show_document? p_table=STANDARDS&p_id=9735

A-weighted average sound level	85	90	92	95	97	100	105	110	115
OSHA PEL (h/day)	16	8	6	4	3	2	1	0.5	0.25
NIOSH recommendation (h/day)	8	2.52	1.59	0.79	0.5	0.25			

Here t_n is the actual exposure time and T_n is the permissible time at a given sound level. If the variations in noise level involve maxima at intervals of 1 second or less, it is considered continuous. However, exposure to impulsive or impact noise should not exceed 140 dB peak sound level. For a working lifetime of 35 years, 15–25% (or average of 20%) of workers may still develop disabling hearing loss.

B.   National Institute for Occupational Safety and Health (NIOSH) Recommendation

Experiments show that the total amount of energy reaching the cochlea is the main cause of hearing loss when the exposure is continuous. Since increasing the sound level by 3 dB doubles the sound intensity, we can reduce the exposure time by half to maintain constant energy. To protect workers in workplace, NIOSH recommends the "Recommended Exposure Limit (REL)" for occupational noise exposure at 85 dB 8-hr Time-Weighted Average (TWA) with 3-dB exchange rate, listed in Table 10.2. The recommended exposure time T_{REL} decreases by a factor of 2 for every 3 dB increment, i.e.

$$T_{\text{REL}}(L) = 8 \times 2^{(85-L)/3} \text{ [hours]}. \tag{10.5}$$

The 3-dB exchange rate comes from the experimental observation that injury is a consequence of the total energy deposited in cochlear tissues. The recommended working hours obey the same "fraction of allowed dose formula" as that of Eq. (10.4). For example, a worker exposed to sound level of 97 dB for 1 h, he/she can work in the environment of 92 dB for additional 4 hours to reach the maximum allowed exposure based on the OSHA rule. However, based on the NIOSH recommendation of Eq. (10.5), the worker's REL is $T_{\text{REL}} = 8 \times 2^{(85-97)/3} = 0.5$ h or 30 min at 97-dB environment. If the worker had worked for 1 hour, the noise dose would have reached 200%. The factory should therefore reduce its noise environment for its workers. The NIOSH REL does not apply to general environmental or recreational noise outside the workplace. The additional sound sources encountered at restaurants, bars, gyms, fashion stores, movies, concerts, or sporting events, or through activities such as listening to loud music, playing music, recreational hunting, and power tool use, etc. may increase the overall risk for hearing loss.

Based on an estimated 40 year working lifetime, the NIOSH recommendation at 85 dB will cause about 5–15% (or average 8%) NIHL, while OSHA regulation of 90 dBA 8-hr-TWA causes about 20% NIHL. In 2009, the European Commission ratified rules mandating a default maximum volume of MP3 players to be less than 80 dB. At this level, some individuals (5% of population) may still develop a significant hearing loss.

★ Example 10.7: What is the worker's permissible exposure limit if he/she works at 100-dB for 1 hour and 95 dB for additional 1.5 hours?

Answer: Based on Table 10.2, the PEL is 2 hours for 100 dB and 4 hours for 95 dB. Using Eq. (10.4), the Fraction of total dose is $\frac{1}{2} + \frac{1.5}{4} = \frac{3.5}{4} = 87.5\%$, i.e. he has used up 85% of the permissible exposure limit. Since the occupational noise is above 85 dB, the employer must institute a *hearing conservation program* that includes regular testing of employees' hearing by qualified professionals — see Occupational noise exposure 29 CFR 1910.95(c).

★ Example 10.8: A worker worked at the 95 dB environment for 1 hour, what is the maximum time for him to continue working in the 105 dB environment under the permissible exposure limit?

Answer: Using Eq. (10.4) and Table 10.1, the Fraction of total dose is $\frac{1}{4} + \frac{t}{1} = 1$, so $t = \frac{3}{4}$ h. The worker's permissible exposure limit in the 105-dB is 45 minutes.

VI. Noise Control and Reduction

Noise reduction relies on the control of sound transmission, reflection, refraction, diffraction, sound absorption, the isolation of sound sources, and elimination of flow turbulence.

A. Sound Isolation and Enclosure of Sound Source

Enclosure of sound source with sound absorbers can reduce noise level up to about 10–11 dB. Heating, ventilation, air-conditioning (HVAC) systems are the most common annoying noise source in homes, offices, classrooms, and concert-halls. HVAC systems have compressors, pulleys, belts and fans to generate and transport heat from one place to another. Each of these components produces noise. Modern HVAC systems have enclosures to reduce noise pollution. At the inlet and outlet, the addition of mufflers can minimize the flow noise. Smooth air transport handling in air ducts can minimize noise generated by the turbulent flow.

B. Sound Absorbers

Absorption of a sound wave requires the conversion of sound energy into heat. One effective way is to use porous materials made of many small fibers or cells, which causes friction between the air molecules in sound transmission. Porous absorbers, such as perforated tiles or acoustic tiles, are more effective against high frequency sounds. Lightweight panels are effective on the absorption of sound wave at low frequencies because they can be set into motion by converting sound energy into internal friction (heat). Perforated panels with sound absorbing materials that can enhance sound absorption at certain resonance frequencies have become basic building materials in industry. Sound absorbers and sound diffusers have become relevant building materials in concert halls and auditoriums. An anechoic room, which can absorb all sound energy impinging on the wall, can be designed based on physics principle.

C.   Walls, Floors, and Ceilings

When a sound wave strikes the wall, it reflects back, and the wall absorbs a small portion. The rest is transmitted through the wall. The fractions of reflection, transmission and absorption depend on the physical properties of the wall, as well as sound frequency and incident angle. One defines the transmitted fraction τ and the transmission loss (TL) as

$$\tau = \frac{W_{\text{transmit}}}{W_{\text{input}}}, \quad \text{and} \quad \text{TL} \equiv 10 \log \frac{1}{\tau} \tag{10.6a}$$

where W_{input} and W_{transmit} are the acoustic powers input onto and transmitted through the wall, respectively. At low frequency, sound transmission loss obeys the *mass law*, i.e. the transmission loss TL_0 at the normal $(0°)$ incident angle is

$$\text{TL}_0 = 10 \log \left(1 + \left(\frac{\pi M f}{\rho v} \right)^2 \right) \approx 20 \log (Mf) - 42, \tag{10.6b}$$

where M is the wall "area mass" in kg/m^2, f is frequency in [Hz] and $\rho v \approx 400$ kg/m^{2}s is the acoustic impedance of air. The averaged transmission loss reduces 5 dB when the incident angle of a sound wave uniformly covers from 0 to $80°$, i.e.

$$\text{TL} = \text{TL}_0 - 5 \text{ dB} \approx 20 \log(Mf) - 47. \tag{10.6c}$$

Figure 10.8(a) shows a schematic drawing of the transmission loss TL with $M = 30$ kg/m^2. Transmission loss for a wall may fall considerably below the predicted "mass law" in the presence of wall resonances and excitation by bending wave at the *critical frequency*. A bending waves is the sound wave that travels inside the wall at the same speed as the sound wave in air. Leakage of the sound wave through holes and cracks can also lower the transmission loss. The transmission loss is small at low frequency.

One typically measures the transmission loss at 16 standard $\frac{1}{3}$-octave bands from 125 to 4,000 Hz, and fits the measured data with a standard curve that resembles the dBA weighting function at equal energy. The nearest "integer" sound level at 500 Hz of the fitted curve is the Sound Transmission Class (STC) of the wall. Table 10.3 lists some STC numbers and their effects on wall insulation. Figure 10.8(b) shows the standard fitting curve at STC = 45, where the sound level at 500 Hz is 45 dB. The STC number is widely used to rate interior partitions, ceilings/floors, doors, windows and exterior wall configurations in USA (see the STC fitting procedure of the American Society for Testing and Materials: https://www.astm.org/Standards/E413.htm). Outside USA, the Sound Reduction Index, defined in ISO 16283: https://www.iso.org/standard/59748.html, is also commonly used. The Department of Housing and Urban Development recommends STC 55 or higher for walls between apartments in non-urban areas.

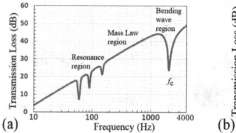 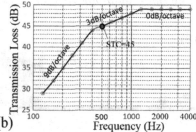

(a) (b)

Figure 10.8: (a) A schematic drawing of the transmission loss for a material with $M = 30$ kg/m^2. (b) A standard curve to fits the measured data of a wall or construction materials at equal sound energy at STC = 45. The STC value reads off from the fitted standard curve at 500 Hz.

Table 10.3: Common STC numbers for building walls and their effects on sound.

STC	Effects
25	Normal speech can be understood quite easily and distinctly through wall
30	Loud speech can be understood fairly well, normal speech heard but not understood
35	Loud speech audible but not intelligible
40	Onset of "privacy"
42	Loud speech audible as a murmur
45	Loud speech not audible; 90% of statistical population not annoyed
50	Very loud musical instruments are faintly heard; 99% of population not annoyed.
60+	Superior soundproofing; most sounds inaudible

D. Barriers (Effect of Diffraction) — Indoors and Outdoors

Barriers is popular in cubicles and as noise control around highways and railroads. In a cubicle, unless the wall is flimsy (see transmission loss), the effect of sound transmission is small compared with those of reflection and diffraction. The sound reflection depends on the wall absorption coefficients, the acoustic properties of the ceiling, the size of the opening above the barrier, and the nature of the walls on the source and received side of the barrier. A highly absorbent ceiling is important in an open-plan office.

Figure 10.9: (a) Geometric drawing of a barrier between a noise source (highway) and a noise protection (residential) area. (b) The attenuation vs the Fresnel number.

Sound barriers that block the direct path from the source to receivers can appreciably reduce noise. The major sound transmission mechanism in barriers is diffraction, which is important for low-frequency sound waves. The attenuation A_{barrier} due to diffraction of sound around a barrier is[12]

$$A_{\text{barrier}} = 20\log \frac{\sqrt{2\pi N}}{\tanh \sqrt{2\pi N}} + 5 \text{ [dB]} \quad \text{with} \quad N = \frac{2(a+b-d)}{\lambda}, \quad (10.7)$$

where N is the Fresnel number, λ the wavelength of sound, and a, b and d are distances in the triangle defined in Fig. 10.9(a). Figure 10.9(b) shows the attenuation vs Fresnel number. A practical limit of attenuation is about 25 dB. A sound barrier is more effective in attenuating high-frequency noise than low frequency noise. Refraction of sound caused by wind or a temperature gradient may lead to additional trans-

---

[12]U.J. Kurze, *JASA*, **55**, 504 (1974); U.J. Kurze and G.S. Anderson, *Appl. Acoust.*, **4**, 35 (1971).

mission of sound over the barrier. To be effective, one needs to build a very high barrier.

★ Example 10.9: What is the Fresnel number for a traffic noise barrier at 20 feet high, 30 feet from the highway and 30 feet from a home for 1-kHz and 100-Hz noise?

Answer: Using the Pythagorean Theorem, we find $a = b = \sqrt{20^2 + 30^2} = 36$ feet. The speed of sound wave at 20°C is $v = 1125$ feet/s. The wavelength are $\lambda = \frac{v}{f} = 1.1$ and 11 feet for $f = 1$-kHz and 100-Hz, respectively. With the parameter $d = 30 + 30 = 60$ m, the Fresnel numbers are $N = 2 \times (36 + 36 - 60)/1.1 = 22$ and $2 \times (36 + 36 - 60)/11 = 2.2$ for 1-kHz and 100-Hz, respectively. The attenuation factors are 27 dB and 16 dB from Fig. 10.9(b). Clearly, the traffic barrier is more effective in blocking the 1-kHz noise than the 100-Hz noise.

E. Aircraft Noise

The total Airlines passenger-miles in the United States increases linearly since 1960. Aircraft noise has become very important in metropolitan cities. Figure 10.3(a) shows that the aircraft noise power has reduced more than 100 times in the 20th century. The aircraft noise measurement is the Perceived Noise Level (PNL) scale. The PNL requires measurements of sound levels in 8 octave bands in the frequency ranges of 20–10,000 Hz. The combination of these data according to a special prescription similar to the A-weighted method arrives at the result in PNdB.[13]

Federal Aviation Regulation (FAR)-Part-36 uses the effective perceived noise level (EPNL) for certification of aircraft noise. It takes the duration and presence of discrete frequency tones into account. The EPNL combines the sound pressure levels in 24 $\frac{1}{3}$-octave bands at nominal mid-band frequencies from 50 Hz to 10 kHz, inclusive, and integrates over the duration of an aircraft flyover. It involves a correction

---

[13]K.D. Kryter, *JASA*, **31**, 1415 (1959), adopted by the ISO with updated procedure in ISO-3891:1978: "Acoustics – Procedure for describing Aircraft Noise heard on the ground." See also https://www.law.cornell.edu/cfr/text/14/part-36.

factor that adds to the PNL when there are discrete tones in the noise spectrum. It also includes a correction obtained by integrating the PNL over a 10-second time interval. Derived from EPNL, the noise-exposure forecast (NEF) takes into account the frequency of the events in a given neighborhood. FAR-Part-36 imposes noise-level limits on aircraft certified after 1969. Stage-1 aircraft, certified before 1969, are being phased out. Modern turbofan jet engines radiate about 0.02% (20 ppm) of its total power as sound [see Fig. 10.3(a)]. This is very much improvement over those used in Boeing 707 or DC8 at the same engine thrust. The NASA quiet engine project aims to reduce noise power further.

F. Shockwave and Supersonic Aircraft

A sonic boom or shockwave is a pressure transient of short duration that occurs during the flyover of an object at a speed exceeding the speed of sound. This sonic boom can occur from supersonic jets. The size of sonic boom depends on the size, the Mach number and the height of the flying object. At ground level, a momentary over/under pressure of 10–100 N/m^2 occurs during the passage of the shockwave (Mach cone) from a supersonic jet. Experiments can observe 20 N/m^2 at 0.2–0.3 s when a Concorde flies over at an altitude of 40,000 ft. The sonic boom covers about 1 mile wide for each 1000 feet of altitude of the aircraft. Regulation of the FAA forbids sonic boom production over land area in the US territory by civilian aircraft.[14] A meteorite or asteroid entering the atmosphere at high speed can also create dangerous shockwave around its path.

G. Ultrasound and Infrasound

Ultrasound is a sound wave with frequencies higher than the audible range, or above 20 kHz. Infrasound refers to a sound wave at frequencies below 16 Hz.

Waterfalls, volcanos, ocean waves, earthquakes, wind, thunder, jet-aircraft, air-conditioning systems, wind turbine and other machinery radiate infrasound of moderate intensity. Headache, coughing, blurred

---

[14]See http://www.gpo.gov/fdsys/pkg/CFR-2003-title14-vol1/xml/ CFR-2003-title14-vol1-part36.xml for further details.

vision and nausea are among the symptoms reported from excess exposure to high levels of infrasound.

Jet engines, high-speed drills, ultrasound cleaning devices, etc. emit ultrasound. The discovery of piezoelectricity in 1880 paves the development of high intensity ultrasound devices for medical and industrial imaging, and for medical surgeries. These procedures are relatively safe, while damage may result as a direct result of acoustic cavitation. There are reports of people experiencing nausea, headache, and changes in blood sugar due to exposure to high intensity ultrasound. There has been no report of adverse effect for SPL below 105–115 dB. Diagnostic ultrasonic exposure is based on the benefits gained from the necessary diagnostic information under the "as low as reasonably practicable" (ALARP) principle. The introduction of intense high frequency noise in the anti-loitering device "the mosquito" for cloud control in public area can affect public health. Fortunately, ultrasonic waves attenuate strongly in air; they do not travel very far from the source.

VII.   Summary

Sound is indispensable in art and communication, but loud sound can be annoying, and this is called noise. Loud sound wave can cause damage to our hearing systems. Two main types of permanent hearing loss are conductive hearing loss and sensorineural hearing loss. Conductive hearing loss comes from defects or damages in the outer or middle ear, preventing sound from passing efficiently to the cochlea in the inner ear. Sensorineural hearing loss arises from damage to the receptor organs of the inner ear. Possible causes of hair cell damage come from prolonged exposure to loud sound, aging, illness, head trauma, toxic medications or genetic defects.

In 1983, the Occupational Safety and Health Administration (OSHA) set up the "Permissible Exposure Level (PEL)" law. The PEL is 90 dBA 8-hr Time-Weighted Average (TWA) with 5 dB exchange rate. The OSHA hearing conservation program requires employers to monitor noise exposure levels in a way that accurately identifies employees exposed to noise at or above 85 dBA 8-hour time-weighted average (TWA). Based on the "energy equivalent principle," the National Institute for Occupational Safety and Health (NIOSH) recommends the "Recommended Exposure Limit (REL)" for occupational noise exposure at 85 dBA 8-hr TWA with 3 dB exchange rate. The World Health Organization (WHO) also recommends 85 dBA 8-hr TWA with 3 dB exchange rate for factories and workplaces. Some people are born with genes that make them more susceptible to loud noises and hearing loss. Avoiding loud sound sources may delay the onset of NIHL.

There has been much progress on technologies in noise control, such as noise reduction by turbulence reduction, noise source enclosure and confinement, etc. Research and development on sound absorption materials, transmission loss and diffraction with sound barriers are important in building construction. Research and development on noise cancellation devices, hearing protection systems, hearing aids, and the cochlear implant continue to be active, and the available devices are getting better.

VIII.   Homework 10

1. The A-weighted sound pressure level $L_p(\text{A})$ in unit of dBA is commonly used in evaluating the environmental noise. The DBA filter function resembles the Fletcher–Munson curve at $L_\text{p} =$ _____ phon.

2. The lower and higher cutoff frequencies of the $\frac{1}{3}$-octave band at the center frequency of 1000 Hz are _____ Hz and _____ Hz, respectively.

3. Based on the *energy equivalent level*, the effect in the noisy environments is identical if the total absorbed acoustic energies are identical. Staying at the loudest stadium (29 September 2014, Arrowhead Stadium in Kansas City) with 142.2 dBA sound level for 1 s is equivalent to stay at 95 dB noisy restaurant for _____ s.

4. Based on the energy equivalent level, staying at 120 dB environment for 10 s is equivalent to staying at 110 dB environment for _____ s.

5. Based on the energy equivalent level, staying at 110 dB environment for 10 s is equivalent to staying at 90 dB environment for _____ s.

6. The day-night equivalent level is to add _____ dB to the energy equivalent level.

7. The NASA quiet engine program aims to produce engines with dBA noise power to engine power at _____ ppm (part per million) (see Fig. 10.2).

8. Data of the sound level for a noise source in octave band is listed in the Table below. What is the overall sound level of the source?

Center Frequency (Hz)	125	250	500	1,000	2,000	4,000	8,000
Sound level (dB)	70	80	90	100	90	90	80

9. The acoustic attenuation of air is proportional to (frequency)^n, where $n =$ _____ (see Sec. 16.III).

10. An *audiogram* is a graph that shows the softest sounds a person can hear at $\frac{1}{3}$ octave from 20 Hz to 16,000 Hz with ISO 389-7:2005 standard. The audiogram of the noise induced hearing loss (NIHL) has a dip at _____ Hz.

11. In OSHA regulation, if a worker worked at the 90 dBA environment for 1 h, how long can he work at the 100 dBA environment so that he does not violate the PEL?

12. The Occupational Safety and Health Administration (OSHA) set the Permissible Exposure Level (PEL) rule in 1983. The OSHA act of 1970 set PEL at _____ dBA as an 8-hour time-weighted averaged (TWA) with _____ dB exchange rate.

13. The National Institute for Occupational Safety and Health (NIOSH) *recommends* the "Recommended Exposure Limit (REL)" for occupational noise exposure to be _____ dBA as

an 8-hr Time-Weighted Average (TWA) with _____ dB exchange rate.

14. Under the NIOSH recommendation, if a worker worked at the 90 dBA environment for 1 h, how long can he work at the 95 dBA environment so that he is within the limit of the REL?

15. WHO recommends an ambient noise levels below _____ dBA for optimum sleeping condition.

16. In 2009, the European Commission has ratified rules mandating a default maximum volume of MP3 players to less than _____ dB.

17. The audiogram of a retired factory worker (age 55) is shown below. Is the person NIHL or AHL?

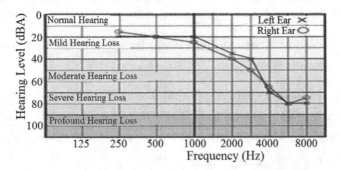

18. A 10 m high barrier is located at mid-way between a highway and a house by 30 m apart. The wavelength to give the Fresnel number 10 is _____ m. This corresponds to frequency of sound less than _____ Hz may have little attenuation. (The speed of sound is 343 m/s.)

19. Find the center frequencies of the $\frac{1}{3}$-octave band from 125 Hz to 4000 Hz that are commonly used in measuring the transmission loss of a wall.

Appendices

Appendix A: Newton's Law of Motion and Bernoulli Principle

(1) Newton's Law of Motion

Newton's laws of motion are the foundation of classical mechanics on the response of a body subjected to external forces. These laws are

1. First law: In an inertial reference frame, an object either is at rest or moves at a constant velocity, unless acted on by a force.

2. Second law: The acceleration of a body is proportional to the net force acting on the body, and inversely proportional to its mass: $F = ma$, where F is the net force acting on the object, and m and a are the mass and acceleration of the object. The direction of acceleration is the same as that of the net force. Newton also stated his 2nd law as "the net force equals the rate of change of momentum mv."

$$F = ma = m\frac{\Delta v}{\Delta t} = \frac{\Delta(mv)}{\Delta t} = \frac{\text{change of momentum}}{\text{time interval for the change}}. \quad (A.1)$$

   We use Newton's second law to derive the speed of wave propagation in a medium.

3. Third law: When one body exerts a force on a second body, the second body will exert a force equal in magnitude and opposite in direction to the first body.

The three laws of motion were first compiled by Isaac Newton in his *Philosophiæ Naturalis Principia Mathematica* (Mathematical Principles of Natural Philosophy), published in 1687. In this work, Newton also discovered the universal law of gravitation describing the force of attraction between two masses:

$$F = G\frac{m_1 m_2}{r^2}, \tag{A.2}$$

where $G = 6.67 \times 10^{-11}$ [N $\cdot$ m^2/kg^2] is called the universal gravitational constant, m_1 and m_2 are masses of two objects, and r is the distance between these two masses. The gravitational acceleration on Earth's surface is $g = G\frac{m_E}{R_E^2} = 9.80$ m/s^2, where $m_E = 5.97 \times 10^{24}$ kg is the mass of the Earth, and $R_E = 6.36 \times 10^6$ m is the mean radius of the Earth.

(2) Bernoulli Principle

When fluid (including airflow) moves in a streamline without turbulence and drag force, the fluid flow is laminar. Figure A.1(a) shows a schematic drawing of a laminar flow. The Bernoulli principle is the *work-energy principle* or conservation of energy:

$$p_1 + \frac{1}{2}\rho v_1^2 + \rho g y_1 = p_2 + \frac{1}{2}\rho v_2^2 + \rho g y_2, \tag{A.3}$$

where p is the pressure, ρ is the fluid density, v is the speed of fluid flow, y is the height, and $g = 9.8$ m/s^2 is the gravitational acceleration constant on Earth's surface. The indices 1 or 2 represent these physics quantities at locations 1 and 2, respectively. In a laminar flow at a constant fluid density, the equation of continuity is $A_1 v_1 = A_2 v_2$, where A_1 and A_2 are cross-sectional areas at different locations. Bernoulli principle is the working principle of vocal fold vibration and reeds of musical instruments. Figure A.1(b) schematically shows the lift force of an airfoil due to the Bernoulli principle. A slower wind speed at the bottom of the airfoil has a higher pressure, which pushes the airfoil upward to provide a lift force.

In fluid flow, the power in moving fluid is the product of the pressure and the volume flow rate:

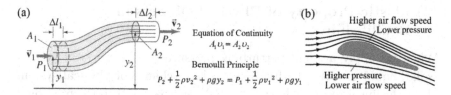

Figure A.1: (a) The laminar flow of air or liquid obeys the Bernoulli principle, where the work done by the pressure is equal to the change of mechanical energy. (b) A schematic drawing of lift force of airfoil based on the Bernoulli principle. The air flows faster through the upper surface of the airfoil, and thus a smaller pressure that provides a lift force of the airfoil. Similar principle is applicable to the vibration of reeds in musical instruments.

$$ W = p \frac{\Delta V}{\Delta t}, \qquad (A.4) $$

where p is the pressure that moves the fluid, $\frac{\Delta V}{\Delta t} = Av$ is the volume flow rate, A is the area, and v is the flow speed of the fluid.

★ Example A.1: A professional player playing a trumpet with over-pressure 10 kPa and the air volume flow rate 200 cm^3/s. The sound pressure level measured at 1 m from the trumpet is 80 dB. What is the power in playing the trumpet?

Answer: The power is the product of pressure and volume flow rate, i.e.

$$ W = (10{,}000 \text{ Pa}) \times (200 \times 10^{-6} \text{ m}^3/\text{s}) = 2.0 \text{ W}. $$

The sound intensity at 1 m from the trumpet is $I = I_0 \times 10^{80/10} = 10^{-4}$ W/m^2, where the reference intensity is $I_0 = 10^{-12}$ W/m^2. The sound power is $W_{\text{Sound}} = I \times (4\pi r^2) = 0.00126$ W. The efficiency in sound production is $0.00126/2.0 = 0.063\%$.

★ Example A.2: In loud singing at 120 dB, the pressure is $p = 4{,}000$ N/m^2 with flow rate 400 cm^3/s, what is the power involved in loud singing?

Answer: The power is the product of pressure and volume flow rate, i.e.

$$ W = (4{,}000 \text{ Pa}) \times (400 \times 10^{-6} \text{ m}^3/\text{s}) = 1.6 \text{ W}. $$

(3) Elastic Property of Physical Objects in Equilibrium

When we apply a small perturbing pressure p on a physical system (solid, liquid or gas), the fractional volume change of physical system obeys the linear response law: $p \equiv -B\frac{\Delta V}{V}$, where $\Delta V/V$ is the fractional volume change, B is the bulk-modulus, the "$-$" sign provides a positive number for the bulk-modulus. Within the "elastic limit," the elastic property is

$$(\text{stress}) = (\text{elastic modulus}) \times (\text{strain}), \qquad (A.5)$$

where pressure or the *force per unit area* represents the stress, fractional volume change is the strain, and the bulk modulus is the elastic modulus. The "tensile strength" is the "maximum stress" before the breaking point of the physical system.

Appendix B: Basic Mathematics

(1) Sinusoidal Functions (Sine and Cosine)

The solution of simple harmonic motion is the sine-cosine functions in Eq. (2.3). Mathematicians set up the sine and cosine functions in around 2000–1000 BC. Figure A.2 shows the triangle ABC, the angle θ, the radius r or the hypotenuse of the triangle, the adjacent and opposite sides, a and b, and the subtended arc length l of the angle θ in the circle.

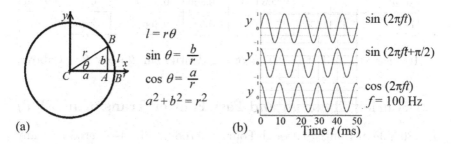

Figure A.2: (a) Definition of angle and the trigonometry functions. (b) (Top) The sine function $\sin(2\pi ft)$ vs time in [ms] of Simple Harmonic Motion (SHM) at $f = 100$ Hz; (Middle) the sine function $\sin(2\pi ft + \frac{\pi}{2})$ with phase shifted by $\pi/2$ or $90°$ at $f = 100$ Hz; (Bottom) the cosine function $\cos(2\pi ft)$ at $f = 100$ Hz. The $\pi/2$ phase shifted sine function is equal to the cosine function.

The angle θ is the subtended arc length divided by the radius. The angle defined this way is length/length, i.e. no dimension, called "radian" or [rad]. The sine and cosine functions are

$$\theta = \frac{l}{r}, \qquad \sin \theta = \frac{b}{r}, \qquad \cos \theta = \frac{a}{r}. \qquad (A.6)$$

The arc length of one complete circle is $2\pi r$, i.e. the angle for one complete circle is 2π [rad]. In human history, $360°$ angle was set to a complete circle before 250 BC, i.e. $360° = 2\pi$ rad, or 1 rad $= 57.296°$. The *Pythagorean theorem* of a right triangle is $r^2 = a^2 + b^2$, or $\sin^2 \theta + \cos^2 \theta = 1$. Figure A.2(b) shows the sine and cosine functions.

The solution of the SHM in Eq. (2.3) is $y = A\sin(2\pi ft + \varphi)$, which has a single frequency f. Sound waves coming from realistic musical instruments are superposition of many frequencies. When the sound wave propagates through the air, the motion of air molecules is a super-position of sine functions at these frequencies.

★ Example A.3: List the angle in degree and radian transformation. Find their corresponding values of sine and cosine functions.

Answer:

θ (deg)	0	30	60	90	120	150	180	210	240
θ (radian)	0	$\pi/6$	$\pi/3$	$\pi/2$	$2\pi/3$	$5\pi/6$	π	$7\pi/6$	$4\pi/3$
$\sin\theta$	0	0.5	0.866	1	0.866	0.5	0	-0.5	-0.866
$\cos\theta$	1	0.866	0.5	0	-0.5	-0.866	-1	-0.866	-0.5

It is easy to verify the Pythagorean identity $\sin^2\theta + \cos^2\theta = 1$ above.

(2) Fourier Theorem and Fast Fourier Transform (FFT)

In 1807, Jean-Baptiste Joseph Fourier (1768–1830) discovered that any continuous function of time $F(t)$ can be represented by a superposition of sinusoidal functions in frequency space with Fourier amplitude $G(f)$ at frequency f. This is Fourier transform. Fourier transform of a time function produces a "Fourier spectrum." In particular, if a function $F(t)$ is periodic with period T_0, its Fourier spectrum has a fundamental frequency $f_0 = 1/T_0$, and its harmonics are frequencies at $2f_0, 3f_0, 4f_0, \ldots$ (see Fig. 5.5), or integer multiples of the fundamental frequency. Examples of periodic Fourier spectrum are

[1] Periodic Triangular wave: odd harmonics with amplitudes $|A_n| = A_1/n^2$, $n = 1, 3, 5 \ldots$
[2] Periodic Square wave: odd harmonics with amplitudes $|A_n| = A_1/n$, $n = 1, 3, 5 \ldots$
[3] Periodic Sawtooth wave: all harmonics with amplitude $|A_n| = A_1/n$, $n = 1, 2, 3, 4, 5 \ldots$
where A_n is amplitude of the nth harmonic (mode) at frequency $f_n = nf_0$.

In LPCM digitization, the time function $F(t)$ is conveniently repre-sented by discrete samples, $F(t_1), F(t_1 + \Delta t_s), F(t_1 + 2\Delta t_s), \ldots$ at the sampling time-step Δt_s. The Fast Fourier Transform (FFT) is an algorithm that can efficiently calculate the spectrum of a "discrete digitized data sequence" in finite time-window Δt. The Society of Industrial and Applied Mathematics named the FFT algorithm, invented by James Cooley of the IBM and Princeton University and AT&T Bell Laboratories in 1965, as one of top 10 algorithms of the 20th century. The FFT algorithm is available in all mathematical and statistical computational packages, including Excel, MATLAB, LabVIEW, voice synthesizer software, etc. The frequency resolution of Fourier spectrum Δf depends on the FFT time-window Δt according to the "uncertainty principle"

$$\Delta f \Delta t \geq K = \frac{1}{4\pi}. \tag{A.7}$$

In order to achieve frequency resolution of $\Delta f = 1$ Hz, the FFT time window must be larger than $\Delta t \geq \frac{1}{4\pi(\Delta f)} \approx 80$ ms. Figure 4.7 in Chapter 4 shows example of the frequency spectrum at FFT time window of $\Delta t = 20$, 100 and 500 ms, respectively. Note that the frequency resolution improves as the sampling-window increases, until reaching the intrinsic limit of the sound source.

The sampling bandwidth or the sampling rate of the LPCM is $R = 1/\Delta t_s$. The Fourier spectrum covers frequency up to a maximum frequency $f_{max} = R/2$ (Nyquist theorem). Since the frequency range of human ears is up to 20 kHz, CD technology adopts the sampling rate of $R = 44.1$ kHz, i.e. the sampling time-step is $\Delta t_s = 22.67$ μs. Telephone technology uses an 8 kHz sampling rate, or a sampling time of $\Delta t_s = 125$ μs. The sampling rate beyond GHz is available in digital electronics.

(3) Broadband Source, Resonance, Decay Rate vs the Resonance Q-factor

The center frequency and bandwidth of a broadband source with cutoff frequencies f_1 and f_2 are

$$f_c = \sqrt{f_1 f_2} \quad \text{and} \quad \Delta f = f_2 - f_1, \tag{A.8}$$

where f_1 and f_2 are lower and upper cutoff frequencies that the response function is $\frac{1}{2}$ of the maximum. The center frequency is the geometric *average* of cutoff frequencies because the logarithmic nature of our pitch perception. i.e. $\log f_c = \frac{1}{2}(\log f_1 + \log f_2)$. From Eq. (A.8), we find

$$f_1 = \frac{1}{2}\left(-\Delta f + \sqrt{4f_c^2 + \Delta f^2}\right) \quad \text{and} \quad f_2 = \frac{1}{2}\left(+\Delta f + \sqrt{4f_c^2 + \Delta f^2}\right)$$

$$(A.9)$$

The octave band centered at f_c will have the lower and higher cutoff frequencies at $f_1 = 2^{-1/2}f_c$ and $f_2 = 2^{+1/2}f_c$, that satisfies Eq. (A.8) with $\Delta f = f_2 - f_1$. Similarly, the $\frac{1}{3}$ octave band has the lower and higher cutoff frequencies at $f_1 = 2^{-1/6}f_c$ and $f_2 = 2^{+1/6}f_c$.

The frequency response of a resonance in intensity is

$$I(f) = \frac{I_0}{1 + Q^2 \left(\frac{f}{f_0} - \frac{f_0}{f}\right)^2} \approx \frac{I_0}{1 + 4\frac{Q^2}{f_0^2}(f - f_0)^2}, \qquad (A.10)$$

where f_0 is the resonance frequency and Q is the resonance quality-factor or Q-factor (see Fig. 2.13). The upper and the lower cutoff frequencies of the resonance width, defined as $\frac{1}{2}$ of the maximum intensity, are $f_2 = f_0 + \frac{1}{2}\frac{f_0}{Q}$ and $f_1 = f_0 - \frac{1}{2}\frac{f_0}{Q}$ with $f_1 f_2 = f_0^2(1 - \frac{1}{4Q^2}) \cong f_0^2$ for $Q \gg 1$. The bandwidth is $\Delta f \equiv f_2 - f_1 = \frac{f_0}{Q}$, and the Quality factor is

$$Q = f_0/\Delta f. \qquad (A.11)$$

Expressing the intensity in Eq. (A.10) as a time function, we find

$$I(t) = I_0 e^{-2\pi f_0 t/Q} = I_0 e^{-2\pi(\Delta f)t} = I_0 e^{-t/\tau}, \qquad (A.12)$$

where I_0 is the initial intensity, $e = 2.71828\ldots$ is the natural number. The intensity decay time is

$$\tau = \frac{Q}{2\pi f_0} = \frac{1}{2\pi \Delta f} \quad \text{or} \quad Q = 2\pi f_0 \tau \qquad (A.13)$$

(4) Root-Mean-Square (RMS)

Given a set of numbers $\{x_i, i = 1, 2, \ldots, N\}$. The average and rms values of the set are respectively

$$\bar{x} = \frac{1}{N}\sum_{i=1}^{N} x_i, \qquad \text{RMS} = \sqrt{\frac{1}{N}\sum_{i=1}^{N} x_i^2} \qquad (A.14)$$

Sinusoidal function oscillates about an average value. Since physical quantities such as the power, intensity, depend on the square of the sinusoidal function. The root-mean-square value of these physical quantities becomes the relevant measure. The average of a sinusoidal function $y = A\sin(2\pi f t + \varphi)$ is $\bar{y} = 0$, and its rms is $\frac{A}{\sqrt{2}}$, independent of the frequency f. The average power in an AC circuit is $I_{rms}^2 R$, where I_{rms} is the rms value of the electric current and R is the resistance of a resistor. The sound wave is composed of sinusoidal oscillations of pressure wave, and the intensity of a sound wave is proportional to $\Delta p_{rms}^2 = \frac{1}{2}\Delta p_{max}^2$. The factor $\frac{1}{2}$ is the average of the square of the oscillating physical quantity.

★ Example A.7: Consider a sinusoidal function $y = A\sin(2\pi t/T)$, where T is the period. Sample the function at 8 equal time interval of the period and find the rms value of the sine function.

Answer: We sample the function in one period for 8 points at $(t = 0, T/8, 2T/8, 3T/8, 4T/8, 5T/8, 6T/8, 7T/8)$ to obtain function values at $y = (0, A\sin(\pi/4), A\sin(\pi/2), A\sin(3\pi/4), A\sin(\pi),$ $A\sin(5\pi/8), A\sin(6\pi/8), A\sin(7\pi/8)) = (0, 0.707A, A, 0.707A, 0,$ $-0.707A, -A, -0.707A)$. This mean-square is $\frac{1}{8}[0^2 + (0.707A)^2 + A^2$ $+ (0.707A)^2 + 0^2 + (-0.707A)^2 + (-A)^2 + (-0.707A)^2] = \frac{1}{2}A^2$. The root-mean-square is $\sqrt{(\frac{1}{2})A^2} = 0.71A$.

(5) Logarithmic and Exponential (Power) Functions

Exponential and Logarithmic Functions Base the Natural Number e = 2.71828...

Since the acoustic energy absorption rate is proportional to the total acoustic energy in a room, the decay of acoustic energy obeys the exponential decay law: $U = U_0 e^{-t/\tau}$, where U_0 is the initial acoustic energy in the room, τ is the "decay time constant," and $e = 2.71828...$ is the natural number, sometimes called Euler's number [see also Eq. (A.13)]. The Logarithmic function associated with the natural number is the *natural logarithm* denoted by "Ln" or "ln". General operation of the natural logarithmic function is similar to the logarithm based 10, i.e.

$$y = e^x \longleftrightarrow x = \ln(y) \qquad (A.15)$$

Exponential and Logarithmic Functions Base 10

Measurements in Human societies commonly use based-10 system. We express our decimal number in power of 10: For example, one hundred is 10^2; one thousand is 10^3; one million is 10^6; and one billion is 10^9; etc. The exponential function based 10 has the properties:

$$10^x \times 10^y = 10^{x+y}; \quad \frac{10^x}{10^y} = 10^{x-y}; \quad 10^0 = 1. \tag{A.16}$$

Some examples of exponent operations are

$$(10^2)(10^3) = 10^{2+3} = 10^5; \frac{10^2}{10^3} = 10^{2-3} = 10^{-1} = \frac{1}{10};$$

$$(10^{-2})(10^{-3}) = 10^{-2+(-3)} = 10^{-5}.$$

The Logarithmic function is the inverse of the exponential function:

$$y = 10^x \longleftrightarrow x = \text{Log}(y) \equiv \log_{10}(y). \tag{A.17}$$

We commonly use notations "Log," "Log10," "Log$_{10}$," "log," "log10," or "log$_{10}$" for the base-10 logarithm function. In this book, we use "Log" or "log" exclusively to represent the logarithmic function based 10. Some properties of the Log function are:

$$10^2 = 100 \to 2 = \text{Log}(100); 10^{0.301} = 2 \to 0.301 = \text{Log}(2);$$

$$\text{Log}(y) = 2.3 \to y = 10^{2.3} = 200$$

Let $A = 10^{x_1}$, $B = 10^{x_2}$ then $x_1 = \log A$, $x_2 = \log B$

$$A \times B = 10^{x_1} \times 10^{x_2} = 10^{x_1+x_2}; \ \log(A \times B) = x_1 + x_2 = \log A + \log B$$

$$\frac{A}{B} = \frac{10^{x_1}}{10^{x_2}} = 10^{x_1-x_2}; \qquad\qquad \log\frac{A}{B} = x_1 - x_2 = \log A - \log B$$

$$A^n = (10^{x_1})^n = 10^{nx_1}, \qquad\qquad \log A^n = nx_1 = n\log A$$

These formulas are very useful in the evaluation of log and exponential functions. Some Examples of Log functions are

1. $\text{Log}(4) = \text{Log}(2^2) = 2\text{Log}(2) = 2 \times 0.3010 = 0.6020 \approx 0.6$

2. $\text{Log}(5) = \text{Log}(\frac{10}{2}) = 1 - \text{Log}(2) = 1 - 0.3010 = 0.6990 \approx 0.7$

3. $\text{Log}(6) = \text{Log}(2 \times 3) = \text{Log}(2) + \text{Log}(3) = 0.3010 + 0.4771 = 0.7782$

4. $\text{Log}(8) = \text{Log}(2^3) = 3 \times \text{Log}(2) = 3 \times 0.3010 = 0.9030 \approx 0.9$

5. Evaluate $\text{Log}(9)$ from $\text{Log}(3)$: $\text{Log}(9) = \text{Log}(3^2) = 2\text{Log}(3) = 2 \times 0.4771 = 0.9542$.

6. Table A.1 lists some log function values.

Table A.1: Some values of commonly used Log function.

y	Log(y)	y	Log(y)	y	Log(y)	y	Log(y)	y	Log(y)
1	0.0000	5	0.6990	9	0.9542	13	1.1139	17	1.2304
2	0.3010	6	0.7782	10	1.0000	14	1.1461	18	1.2553
3	0.4771	7	0.8451	11	1.0414	15	1.1761	19	1.2788
4	0.6021	8	0.9031	12	1.0792	16	1.2041	20	1.3010

★ Example A.4: Find the sound intensity of a sound at an intensity level of 85 dB.

Answer: To find the intensity level for $\text{SIL} = 85$ dB, you equate $10\,\text{Log}(I/I_0) = 85$.

(1) Divide both sides by 10 to obtain $\text{Log}(I/I_0) = 8.5$;
(2) Using Eq. (A.17), we find $I/I_0 = 10^{8.5}$; or $I = I_0 \times 10^{8.5} = 10^{-12} \times 10^{8.5} = 10^{-3.5} = 0.000316$ W/m^2.

★ Example A.5: The SPL of a sound wave is 94 dB. What is its rms sound pressure?

Answer: The SPL of the sound wave is $\text{SPL} = 20\log \frac{p}{p_0} = 94$ dB, where $p_0 = 20 \times 10^{-6}$ N/m^2 is the rms reference level, i.e. $\log \frac{p}{p_0} = \frac{94}{20}$. The solution is $p = 10^{\frac{94}{20}} p_0 = 1.0$ N/m^2. This means that the 94 dB sound level is equivalent to an rms sound pressure of 1 N/m^2.

★ Example A.6: Find the SPL of a sound wave with rms pressure of 1.0 N/m^2.

Answer: The SPL of the sound wave is $\text{SPL} = 20\log \frac{1.0\,\text{N/m}^2}{20 \times 10^{-6}\,\text{N/m}^2} = 20\log(5 \times 10^4) = 94$ dB.

Exponential and Logarithmic Functions Base 2

The exponential function and its inverse function can have different bases. For example, our perception of pitch is a base-2 system, called octave. We use the exponential function base 2 to represent the Greenwood-CF-formula for easy description of octave bands in Eq. (3.1). The exponential and logarithmic functions of base-2 system are inverse function to each other:

$$y = 2^x, \quad \longleftrightarrow \quad x = \text{Log}_2 y \tag{A.18}$$

For convenience, we use Log to represent base-10 logarithmic function. The conversion from Log base 10 to the natural-logarithm and to the logarithm base 2 are

$$\text{Log}(x) = \text{Log}(e) \times \text{Ln}(x) = 0.4343 \times \text{Ln}(x); \qquad \text{Ln}(x) = \text{Log}(x)/0.4343 \tag{A.19a}$$

$$\text{Log}(x) = \text{Log}(2) \times \text{Log}_2(x) = 0.301 \times \text{Log}_2(x); \quad \text{Log}_2(x) = \text{Log}(x)/0.301 \tag{A.19b}$$

(6) Solution of Quadratic Equation

Quadratic equation $ax^2 + bx + c = 0$ has 2 solutions:

$$x_1 = \frac{-b + \sqrt{b^2 + 4ac}}{2a} \quad \text{and} \quad x_2 = \frac{-b - \sqrt{b^2 + 4ac}}{2a} \tag{A.20}$$

We can rewrite Eq. (A.8) as $f_1^2 + \Delta f f_1 - f_c^2 = 0$. Using Eq. (A.20) with $a = 1$, $b = \Delta f$, and $c = -f_c^2$, we arrive at the solutions in Eq. (A.9).

Appendix C: Breathing, Lung Capacities, Airflow and Subglottal Pressure

Speech and singing rely on the flow of air through the vocal box. The reservoir of air is our lung. The total lung capacity is about 6 liters and 5 liters for the average human male and female, respectively (see Fig. A.3). The normal respiratory rate is 30–60 breaths per minutes in children and reduces to 12–20 breaths per minute in adults. The typical respiratory volume, called tidal volume, is about 500 cm^3. The vital capacity is the volume available for singing. The change of volume and the subglottal pressure during a speech is shown in Fig. A.3(c).

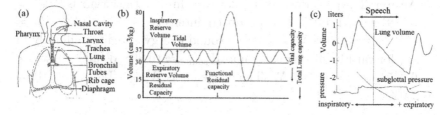

Figure A.3: (a) Schematic model of human lung. (b) Typical lung function (see Table A.2). (c) Lung volume change and subglottal pressure change during a speech (see Ref. [31]).

Table A.2: Typical respiratory information of human lung.

Category of functions	Volume (cm^3)	
	In men	In women
Inspiratory reserve volume (IRV)	3,100	1,900
Tidal volume (TV)	500	500
Expiratory reserve volume (ERV)	1,200	700
Vital capacity (VC = IRV + TV + ERV)	4,600	3,100
Residual volume (RV)	1,200	1,100
Functional Residual capacity (FRC = ERV + RV)	2,400	1,800
Total lung capacity (TLC = VC + RV)	5,800	4,200

Subglottal Pressure

The subglottal pressure is the overpressure of air in the trachea below the glottis. During normal quiet breathing, the subglottal pressure is about $p_s = 100$ N/m^2, and about $\Delta V = 500$ cm^3 in 5 s, i.e. the volume flow rate is $\Delta V / \Delta t \approx 100$ cm^3/s. The power of moving this air is $W = p_s(\Delta V / \Delta t) = 0.01$ [W].

★ Example A.7: Measurements show that one needs a subglottal pressure of 3500 Pa with volume flow rate of 400 cm^3/s to produce 116 dB of sound intensity, measured at 3 inches in front of a singer. What is the efficiency in loud singing?

Answer: The power needed to move air through the larynx is $W_{\text{lung}} = p_s\left(\frac{\Delta V}{\Delta t}\right) = 1.4$ W. The produced intensity is $I = 10^{116/10}I_0 = 10^{11.6-12} = 0.398$ W/m^2. If the sound energy spreads uniformly in the forward hemisphere, we find $W_{\text{sound}} = (0.398 \text{ W/m}^2) \times (2\pi(3 \times 0.0254 \text{ m})^2) = 0.0145$ W. The efficiency is $W_{\text{sound}}/W_{\text{lung}} = 0.010 = 1.0\%$.

Appendix D: Equations and Formula (Cheat-Sheet) in Physics of Sound

Density ($\rho = M/V = $ Mass/Volume)

   Water $= 1000$ kg/m^3;   Nylon $= 1140$ kg/m^3;   Steel $= 7700$ kg/m^3; air $= 1.292$ kg/m^3

Newton's 2nd Law of motion: $F = ma$;
The unit of force is [N] $=$ [Newton] $=$ [kg m/s^2]

Newton's Universal gravitational force: $F = Gm_1m_2/r^2$;
$G = 6.67 \times 10^{-11}$ m^3/(kg s^2)

Gravitational acceleration on Earth's surface: $g = 9.80$ m/s^2

Hooke's Law: Restoring force $F = -ky$; $k = $ spring constant;
$y = $ displacement from equilibrium

Periodic motion: $f = 1/T$;   frequency f [Hz];   period T [s];
Amplitude A; $\text{PE}_{\text{max}} = (\frac{1}{2})kA^2$

Resonance frequency: $f = (1/2\pi)\sqrt{(k/m)}$ for mass on a spring;
$f = (1/2\pi)\sqrt{(g/l)}$ for pendulum

Resonance bandwidth $\Delta f = f_0/Q$, where Q is the quality-factor and f_0 is the resonance frequency

Kinetic energy: KE $= (\frac{1}{2})mv^2$;
Potential energy: $\text{PE}_{\text{grav}} = mgh$ & $\text{PE}_{\text{spring}} = (\frac{1}{2})ky^2$

Work: Work $\equiv$ (Force) $\times$ (displacement) $= F \times d$;
Unit: 1 [N][m] $= 1$ [J] $=$ [Joule]

Work-energy principle: $\text{KE}_1 + \text{PE}_1 + $ Work $= \text{KE}_2 + \text{PE}_2$, where Work is the work done on the system between the positions 1 and 2.

Conservation of energy: If Work $= 0$, then KE$+$PE $=$ constant $= \text{E}_{\text{total}}$;
$(\frac{1}{2})mv^2 + (\frac{1}{2})ky^2 = (\frac{1}{2})kA^2$, where A is the amplitude of SHM.

Power $=$ Energy/time;   Unit of Power is [W] $=$ [Watt] $=$ [J/s]

Intensity $=$ Power/Area;   Power $W = $ Intensity $\times$ Area $= I \times A$

Unit: Power W (W), Intensity I (W/m^2) and Area(m^2)

Intensity spreads onto a sphere at distance r from a source of power W; $I = W/(4\pi r^2)$

Intensity spreads onto a hemisphere at distance r from a source of power W; $I = W/(2\pi r^2)$

Pressure = Force/Area;   Work = pressure $\times \Delta$(Volume);
Power = pressure $\times$ (Volume-flow-rate)

Wave speed: $v = \lambda \times f$

— sound wave in air: $v = 331.3 + 0.6\ t[°C]$ m/s
$$v = 343.3 \text{ m/s at } t = 20°C$$

— wave on string: $v = \sqrt{\frac{F_T}{M/L}}$, where F_T is the tension, M and L are mass and length of string

Wave propagation:

— reflection: $(\theta_{\text{incident}} = \theta_{\text{reflection}})$;

— refraction: wave change propagation direction when the speed of wave is different in medium

— diffraction: $a\sin\theta \sim \lambda$; where a is the size of obstructions, λ the wavelength, and θ the spreading angle

— interference:

due to path length difference, $(d_2 - d_1)$ or

due to frequency difference, $f_2 \approx f_1$, $f_{\text{fused tone}} = (\frac{1}{2})(f_2 + f_1)$; $f_{\text{beats}} = |f_2 - f_1|$

Modulation of tone f_2 by tone f_1 with $f_2 \gg f_1$ creates sidebands at $f_2 \pm f_1$

Doppler Effect: $f' = f_s \frac{v \pm v_o}{v \mp v_s}$; v = speed of sound;
v_o = speed of observer; v_s = speed of source

Harmonic Series for Standing waves: harmonic number n

On a string: $\lambda_n = 2L/n$, $f_n = v/\lambda_n = nf_1$, for $n = 1, 2, 3 \ldots$;
$v =$ wave speed on string

Open air-filled pipe: $f_n = v/\lambda_n = n\,(v/2L)$, for $n = 1, 2, 3 \ldots$;
$v =$ sound wave speed in pipe

Closed air-filled pipe: $f_n = v/\lambda_n = n\,(v/4L)$, for $n = 1, 3, 5 \ldots$;
$v =$ sound wave speed in pipe

Sound intensity: $I(\text{W/m}^2)$; $I = p^2/(\rho v) \approx p^2/(400)$; p is the rms pressure of sound wave $[\text{N/m}^2]$

Intensity of incoherent sources: $I = I_1 + I_2 + I_3 + \cdots$

Sound Levels

Power Level: L_W (dB) $= 10\log(W/W_0)$, where $W_0 = 1.0 \times 10^{-12}$ W

Intensity level: SIL (dB): $L_I = 10\log(I/I_0)$, where $I_0 = 1.0 \times 10^{-12}$ W/m^2

Pressure level: SPL (dB): $L_p = 20\log(p/p_0)$, where $p_0 = 2.0 \times 10^{-5}$ N/m^2

Loudness level L (phons); Loudness (sone) $S = 2^{(L-40)/10}$

Relative level ΔL_I (dB); $\Delta L_I = L_{I_2} - L_{I_1} = 10\log(I_2/I_1)$;

Pitch and Timbre:

Equal temperament: An octave is divided into 12-semitone; each semitone $= 100$ cents.

Semitone: frequency ratio of $2^{1/12} = 1.0595$ or increased by 5.95%;

Diatonic scale: C*D*EF*G*A*BC; $f_n = f_0 \times 2^{n/12}$; $n =$ the number of semitones above f_0

Whole tone: frequency ratio of $(1.0595)^2$ or increased by 12.2%

Conversion of frequency difference Δf at f_0 to "cent" is $x(\text{cent}) = 1730\Delta f/f_0$.

Typical Human Hearing: range of hearing of human ears: 20 Hz to 20 kHz

Equal Loudness Level (ISO 226:2003) revised Fletcher and Munson equal loudness level (L_L) for pure tones:

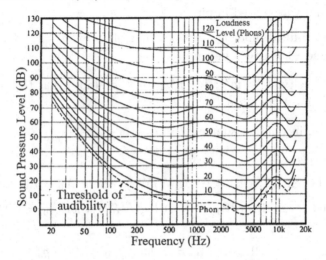

Coulomb's law: $F = kq_1q_2/r^2$; $k = 9 \times 10^9$ Nm^2/C^2; $q_1, q_2 =$ charges, $r =$ distance between charges

Electric current: $I = \Delta Q/\Delta t$: Charge flow rate is electric current in [Ampere] = [Coulomb/second]

Resistors in series: $R_{eq} = R_1 + R_2 + R_3 + \cdots$;
Resistors in parallel: $\frac{1}{R_{eq}} = \frac{1}{R_1} + \frac{1}{R_2} + \frac{1}{R_3} + \cdots$

Ohm's law: $V = IR$; $V =$ electric potential; $R =$ resistance [Ohms]; and $I =$ current [Amperes]

AC voltage: $V_{rms} = V_{max}/\sqrt{2} = 0.707\ V_{max}$;
$\qquad\qquad I_{rms} = I_{max}/\sqrt{2} = 0.707\ I_{max}$;

Ohm's law: $V = IZ$; $V =$ electric potential; $Z =$ reactance [Ohms]; and $I =$ current [Amperes]

$Z_C = 1/(2\pi fC)$; C is the capacitance in [Farad]; and $Z_L = 2\pi fL$; L is the inductance in [henry].

Power of electric circuit: $P = I^2 R = V^2/R = IV$;
$$P_{av} = I_{rms}^2 R = V_{rms}^2/R = IV; \quad P_{av} = \tfrac{1}{2}P_{max}$$

Impedance of RLC circuit: $Z = [R^2 + (Z_L - Z_C)^2]^{1/2}$;
Power factor $= R/[R^2 + (Z_L - Z_C)^2]^{1/2}$.

Resonance frequency of RLC electric circuit: $f_0 = \frac{1}{2\pi\sqrt{LC}}$

Transformers: $\frac{V_s}{V_p} = \frac{N_s}{N_p}$

Digitization: The number of intervals in N-bit digitization is $2^N - 1$.

Dynamic range (DR) $\equiv 20\,\mathrm{Log}\,V/\Delta V$, where ΔV is the voltage resolution.

The DR for N-bit digitization of AC signal is $6.02 \times (N - 1)$ dB.

Standard Prefixes: mega (M) $= 10^6$; kilo (k) $= 10^3$; milli (m) $= 10^{-3}$; micro (μ) $= 10^{-6}$;

Some unit conversion: Temperature °F to °C: $t(°C) = \tfrac{5}{9}[t(°F) - 32]$

1 kg $= 1000$ gm; 1 N $= 1$ kg$\cdot$m/s^2; 1 J $= 1$ N$\cdot$m $= 1$ kg$\cdot$m^2/s^2;
1 s $= 1000$ ms; 1 hour $= 3600$ s;

1 in $= 2.54$ cm; 1 ft $= 0.3048$ m; 1 mile $= 1609$ m; 1 liter $= 1000$ cm^3;

1 Pa $= 1$ N/m^2 $= 10$ dyne/cm^2; 1 cm-H$_2$O $= 9.8$ Pa; 1 lb/in^2 $= 6981$ Pa.

Appendix E: Answer to Homework Problems

Homework 01

1. 29 m/s; 10 gallons; 5.4 h

2. (a) 6 s, 14 s (b) (2 s, 4 s);
   (8 s, 10 s); (c) (4 s, 8 s);
   (d) −2 m/s, +1.0 m/s; (e) 0;
   (f) 9 m

3. 767 mi/h; 1125 ft/s

4. 59 kg; 11000 Pa

5. 2×10^{-10}

6. 21°C

7. 160 J; 100 W
   (2 significant figures)

8. 116 J; 116 J

9. 2.87 m

10. 17.2 m, 0.0172 m

11. Linear; Equilibrium; restoring

12. (a) t_2; t_4; (b) t_5–t_1 or t_4;
    (c) t_1, t_3, t_5

13. Resonance

14. (a) 19.6 N/m; (b) 1.0 Hz;
    (c) 0.63 m/s; 0.098 J

15. 0.35 Hz; 2.8 s

16. Period; frequency

17. "Nearer to"

18. Simple harmonic; frequency;
    amplitude

19. Periodic; fundamental;
    harmonics

20. 0.1 m; 2 m; 100 Hz;
    $2\pi f A = 62.8$ m/s; $f\lambda = 200$ m/s

21. 0.1 m; 2 m; 100 Hz;
    $2\pi f A = 62.8$ m/s; $f\lambda = 200$ m/s

22. Frequencies

23. 50 ms; 50 μs; 17 m; 0.017 m

24. Speed; frequency; 2.4 m; 0.80 m

25. 5 Hz; 100 Hz; 50 kHz

26. 0.05 s; 0.2 ms; 100 μs

27. 0.466 s

28. 10 GHz

29. 125, 250, 500, 750 Hz

30. At 30.2 s after the midnight

Homework 02

1. Interference; constructive;
   destructive

2. Pitch

3. $v = \sqrt{\frac{1.4 \times 8.314 \times 293}{0.02896}} = 343$ m/s,
   agreeing with $331 + 0.6 \times t(°C)$

4. 72 N

5. 280 m/s (2 sig-figure)

6. 4500 m/s (2 sig-figure)

7. Colder; warmer; slower

8. beat

9. 8600 Hz

10. 349 m/s; 1.05 km; Yes, distance
    is about 1 km in 3 s; or 1 mile in
    5 s

11. 3.0 MHz

12. 6.0 N/m^2; 2.0 N/m^2

13. 1 m; (a) constructive;
    (b) destructive

14. 438 Hz; 442 Hz

15. 502 Hz; 506 Hz

16. Doppler; higher; lower

17. 0; 2.24

18. Shock wave; sonic boom

19. 1062 Hz; 945 Hz

20. Standing; nodes; antinodes

21. 130 m/s

22. 1.5 m; 0.75 m; 0.50 m

23. 83.3 Hz

24. 239 m/s

25. 130 cm; 65 cm; 43.3 cm

26. 184 Hz; 368 Hz; 552 Hz

27. 125 Hz

28. Fundamental; Harmonics

29. 0.66 m; 290 m/s; 92 N

30. 3430 Hz

31. 56.3 Hz

32. 0.33 m

33. 1.31 m

34. 35 Hz; 70 Hz; 105 Hz

35. 103 Hz

36. 18 Hz; 53 Hz; 88 Hz

37. 52 Hz

38. 1465 Hz; assuming the speed of sound in He gas at 20°C is 1005 m/s

39. 511 Hz

40. 4.3 m

41. 3300 Hz

42. Antinode

43. Fundamental; Harmonics

44. $Q = 490/(580 - 380) = 2.5$

45. 11 nm (see Example 3.18)

46. 4 W/m^2; 450 W

Homework 03

1. 1726 Hz

2. Acoustic reflex

3. Decibel

4. 3; 6; 23

5. Logarithm

6. $10^{0.3} = 2$; 1,000

7. Amplitude

8. Amplitude; Frequency

9. 200,000; 4,000,000; 20

10. 60 dB; $I \times A = 7 \times 10^{-11}$ W; Energy $= (7 \times 10^{-11}\,\text{W})(30\,\text{s}) = 2.1 \times 10^{-9}$ J

11. 0.006325 N/m^2; 10^{-7} W/m^2

12. 40,000

13. 191 dB

14. 3.16 W/m^2; 358 W; 105 dB; 60 m

15. 63.25 N/m^2; 0.0032 N

16. 32 m

17. 132 dB

18. 20

19. 53 dB; 56 dB

20. 0.1 W; 110 dB; 0.002 W/m^2; 93 dB; 0.89 N/m^2; 0.0005 W/m^2

21. 74 phon

22. 10.5 sone

23. Critical

24. 26%

25. 5%

26. 24.6 dB

27. $f_2 - f_1 = 165$ Hz, $\sqrt{f_1 f_2} = 1000$ Hz; $\therefore f_1 = 921$ Hz; $f_2 = 1086$ Hz

28. $f_2 - f_1 = 78.6$ Hz, $\sqrt{f_1 f_2} = 500$ Hz; $\therefore f_1 = 462.2$ Hz; $f_2 = 540.8$ Hz

29. Binaural

30. Low; high

31. Intensities

32. Haas

33. Middle

34. Low; high

35. Close

36. 0.002 Pa, 1 sone, 0.02 Pa, 4 sone

Homework 04

1. 22.27 Hz; 22,627 Hz

2. 329.6 Hz; 523.3 Hz; 185 Hz; 116.5 Hz

3. $1 + \Delta f/f = 2^{(y/1200)}$, $y = 12$ cents

4. $x = 0.528$; not far from the $x = 0.5$, at the center of the cochlear membrane

5. Speech; Fourier

6. 8 Hz; 1.6 Hz; 0.80 Hz

7. 250 Hz; 500 Hz; 750 Hz

8. 10 ms; 5 ms; 3.33 ms

9. 100 Hz

10. 200 Hz

11. 250 Hz

12. Odd order

13. 300 Hz; 900 Hz; 1500 Hz; 0.333A; 0.2A

14. 1.0 V; 0.667 V; 0.5 V; 2; 3; 4

15. 0.333 V; 0.12 V; 0.0612 V; 3; 5; 7

16. 6 dB/octave; 6 dB/octave; 12 dB/octave

17. 12 dB/octave

18. 580 Hz; 700 Hz; 820 Hz

19. 10 Hz, arising from 200 Hz and 3rd harmonic of the 70 Hz square wave

20. No; Yes

21. Dissonance

22. 261.63 Hz; 392 Hz; 1.4983; small difference from 1.5 of Pythagorean tuning

23. 2.2 Hz; 8.6 cents; 26 Hz; 8.4% of semitone

24. 195.56 Hz; 293.33 Hz; 660 Hz

25. 130.37 Hz; 195.56 Hz; 293.33 Hz

26. 65.19 Hz; 97.78 Hz; 146.67 Hz

27. Major 4th tuning is multiplying 4/3 to obtain the frequencies: E1: 41.25 Hz; D2: 73.33 Hz; G2: 97.78 Hz

28. 233.1 Hz; 116.5 Hz; 87.3 Hz; 58.3 Hz

29. 640 Hz; 960 Hz; 160 Hz

30. Nonlinearity on higher harmonics of piano strings

31. A0: 27.5 Hz; C8: 8372 Hz; miss-tuned 40 cents below A0 is 26.9 Hz within jnd; and 30 cents above C8 is 8518 Hz, which is above jnd.

32. 96 Hz; 165 Hz; 2340 Hz

33. (22.3 Hz, 44.5 Hz); (11300 Hz, 22600 Hz)

34. 35 Hz; 132 Hz; 1102 Hz

35. Looking at Fig. 6.19(c), the semitone line crosses the $\frac{1}{4}$ CB line at 500 Hz.

36. At low frequencies, the $\frac{1}{4}$ CB line is about 25 Hz. At central frequency of about 200 Hz, its whole tone is about 25 Hz away, and thus dissonance.

Homework 05

1. String; membrane; wood or metal bar; pipe; air stream
2. Octave
3. 1,2, 4,5, 7,8, 10,11, ,,; 1,2,3,4, 6,7,8,9, ...
4. Open
5. Brass instruments; woodwind instruments
6. Oscillating airjet or vibrating reed
7. Valves or slides
8. Closed; open; closed conic
9. $\lambda = 0.66$ m for wave on string; $F_T = 59$ N; $\lambda = 0.78$ m for sound wave in air. Typical tensions of violin strings are 90 N down to 40 N for various type of strings.
10. 2.44 m; 82.8 N; 12.5 m; 12.3 m
11. 448 Hz
12. 432 Hz
13. 61.25 Hz; 183.75 Hz; 306.75 Hz
14. Frequency; bell
15. 17.08 cm
16. 16.5 cm
17. 128 Hz
18. 211 Hz; 1.62 m; 1.62 m
19. 207 Hz; 1.62 m; 1.62 m
20. Node

Homework 06

1. 1000 Hz; 4000 Hz
2. Glottal; vocal; mouth
3. Larynx; Vocal folds
4. Formants
5. Spectrogram
6. (a) 375 Hz; 3875 Hz; 4125 Hz; (b) 875 Hz; 2225 Hz; 3000 Hz
7. (b); the fundamental pitch is 200 Hz
8. 200 Hz; 400 Hz; 600 Hz; 800 Hz
9. Phoneme; vowels; consonants
10. Formants
11. 500 Hz; 1500 Hz; 2500 Hz
12. Vowel
13. Frequency; time
14. 120 dB is $I = 1$ W/m^2; power $= IA = 0.002$ W; Power of lung = pressure × (volume flow rate) $= 1.76$ W
15. 2500 Hz; 3000 Hz
16. 0.01 W; 1.6 W
17. 0.073 W
18. The power of sound is 0.073 W at the mouth opening; the power of moving air through larynx is 1.6 W; the efficiency is $0.073\,\text{W}/1.6\,\text{W} = 4.6\%$
19. 165; 12 s
20. 0.029 m; or 2.9 cm
21. $f_1 = 350\,\text{m/s}/(4*0.03\,\text{m}) = 2900$ Hz

Homework 07

1. voltage
2. $87.60
3. $I = 0.183$ A; $Q = I \times \Delta t = 660$ Coulomb
4. 0.6 A; 7.2 W
5. 0.2 A; 0.8 W
6. 288 Ohms

7. 12.5 W

8. 86 W; 172 W

9. 2.9×10^{-6} F

10. 0.326 A; 0 W: ideal capacitors and inductors do not dissipate energy

11. 0.212 A

12. 500 Hz

13. 663 Hz

14. 11 Hz

15. 20

16. 0.353

17. RC $= 1.59 \times 10^{-5}$ s

18. RC $= 3.18 \times 10^{-4}$ s

19. L/R $= 3.18 \times 10^{-5}$ s

20. L/R $= 3.18 \times 10^{-4}$ s

21. 20 dB/decade; 40 dB/octave

22. The rating means that the amplifier provides 20 V to the output. Connecting the 4 Ω speaker gives 11 W to the speaker; and dissipates 22 W to the amplifier.

23. 0.4%; 0.1 W

24. 1,000 Hz; 94 dB

25. $S_V = S_p + 10\,\text{Log}(Z) - 24 = -51$ dBV; The sensitivity is 0.0028 V/Pa; at 100 dB sound input, the pressure is 2 Pa; and thus the output voltage is 5.6 mV.

26. -48 dBV

Homework 08

1. 0.26×10^{-9} W/m^2; 0.12×10^{-4} W/m^2

2. Digital; dynamic

3. Lossless; lossy

4. 11

5. Nyquist

6. Aliasing; above; below

7. 2 channels; 16 bits; 44.1 kHz

8. 1100; 10110; 1000001; 1111111

9. 6; 45; 257; $2^{17} - 1 = 131{,}071$

10. 1; 262143; $\Delta V = 7.63 \times 10^{-5}$ V; 102 dB

11. 1,411,200 bits/s; 5,673,024,000 bits

12. 1.41×10^{-12} m^2/bit; 1.2 μm^2/bit;

13. $(790/635)^2 = 1.5$; $(790/430)^2 = 3.4$;

14. (a) the period is 4 units

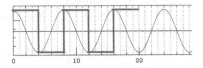

(b) the period is 24 units

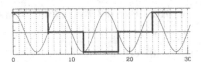

(c) The period becomes infinity or $f = 0$

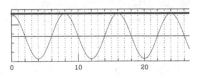

Homework 09

1. (a) 2 m; 171.5 per second;
   (b) 6.4 m^2; (c) RT$_{60}$: Sabine
   0.805 s; Eyring 0.764 s
   (d) d$_{critical}$ = 0.53 m; (e) The
   lowest 3 modes are (1,0,0),
   (0,1,0), (1,1,0) at 42.9 Hz,
   42.9 Hz, and 60.6 Hz;
   (f) Δf = 2.7 Hz;
   (g) f_1 = 42.9 Hz; f_2 = 316 Hz;
   f_3 = 1,265 Hz

2. (a) 3.75 m; 91.5 per second;
   (b) 96 m^2; (c) RT$_{60}$: Sabine
   0.503 s; Eyring 0.423 s
   (d) d$_{critical}$ = 2.34 m; (e) The
   lowest 3 modes are (1,0,0),
   (0,1,0), (1,1,0) at 17.2 Hz,
   17.2 Hz, and 24.3 Hz;
   (f) Δf = 4.37 Hz;
   (g) f_1 = 17.2 Hz; f_2 = 81.6 Hz;
   f_3 = 327 Hz

Homework 10

1. 40 phon

2. $\frac{1}{3}$ octave range
   (891 Hz, 1,122 Hz)

3. 5.2×10^4 s or 14.6 h

4. 100 s

5. 1,000 s

6. 10 dB

7. From Fig. 13.2; the A-weighted
   noise is 10 ppm

8. Similar to example 13.3; the
   overall sound level is 101 dBA

9. n = 2; i.e. f^2

10. 4,000 Hz

11. 1.75 h

12. 90 dBA; 5 dB exchange rate

13. 85 dBA; 3 dB exchange rate

14. 0.48 h

15. 35 dB

16. 80 dB

17. Perhaps both NIHL and AHL;
    no pronounced 4,000 Hz dip

18. λ = 1.2 m; f = 283 Hz; all
    frequency below 283 Hz may
    have little attenuation

19. The center frequencies of the $\frac{1}{3}$
    octave bands are 125, 157, 198,
    250, 315, 397, 500, 630, 794,
    1,000, 1,260, 1,587, 2,000, 2,520,
    3,175, 4,000 Hz

Appendix F: Bibliography (References)

[1] Sheila Tobias, *Overcoming Math Anxiety* (WW Norton, 1995). ISBN-13: 978-0393313079.

[2] **Praat** program: see http://www.fon.hum.uva.nl/praat/

[3] **Audacity** program: https://www.audacityteam.org/

[4] Lord Rayleigh, *The Theory of Sound*, 2nd ed. (Dover Pub., 2nd ed., 1945). ISBN-13: 978-0486602929; https://archive.org/details/theoryofsound02raylrich

[5] Thomas D. Rossing, F. Richard Moore and Paul A. Wheeler, *The Science of Sound*, 3rd ed. (Addison-Wesley, 2002). ISBN: 9780805385656.

[6] Richard E. Berg and David G. Stock, *The Physics of Sound* (Pearson, 3rd ed., 2004). ISBN-13: 978-0131457898.

[7] L. E. Kinsler and A. R. Frey, *Fundamentals of Acoustics*, 2nd ed. (John Wiley, NY, 1962).

[8] Alexander Wood, *Acoustics*, 2nd ed. (New York: Dover, 1966).

[9] H. v. Helmholtz, *Die Lehre von den Tonempfindungen als physiologische Grundlage für die Theorie der Musik* (Braunschweig: 1862). The English translation of the 1877 edition is available as *On the Sensations of Tone* (New York: Dover, 1945).

[10] H. v. Helmholtz, *On the Sensations of Tone* (Dover Pub., 1954). ISBN-13: 978-0486607535.

[11] H. Lamb, *The Dynamical Theory of Sound* (Amazon Digital Services LLC, 2013) https://archive.org/details/dynamicaltheoryo00lambrich, ISBN: 115224471X.

[12] F. Alton Everest and Ken Pohlmann, *Master Handbook of Acoustics* (McGraw-Hill, 6th ed., 2014). ISBN-13: 978-0071841047.

[13] Edmund S. Crelin, *The Human Vocal Tract: Anatomy, function, development, and evolution* (Vantage Press, 1987). ISBN-13: 978-0533069675.

[14] P. B. Denes and E. N. Pinson, *The Speech Chain; the Physics and Biology of Spoken Language* (Waveland Press, 2015). ISBN-13: 978-1478629566.

[15] J. L. Flanagan, *Speech Analysis, Synthesis and Perception* (Springer, Berlin, 1972), 2nd ed. https://link.springer.com/book/10.1007/978-3-662-00849-2

[16] Jan Schnupp, Eli Nelken, and Andrew King, *Auditory Neuroscience (Making Sense of Sound)* (MIT, 2011). ISBN-13: 978-0-262-11318-2; http://auditoryneuroscience.com/book

[17] Springer Handbook of Auditory Research, available at https://link. springer.com/search?facet-series="2506"&facet-content-type="Book"

[18] David Moore, Paul Fuchs, Alan Palmer, Adrian Rees, Christopher Plack, *Oxford Handbook of Auditory Science: The Ear, The Auditory Brain, Hearing.* ISBN-13: 978-0199581412.

[19] Jerry Tobias, Ed. *Foundations of Modern Auditory Theory* (Academic Press, NY, 1970)

[20] Richard M. Warren, *Auditory Perception: An Analysis and Synthesis* (Cambridge University Press, 2008). ISBN-13: 9780521868709.

[21] William A. Yost, *Fundamentals of Hearing, Fifth Edition: An Introduction* (Emerald Group Publishing; 5th Edition, October 2, 2006). ISBN-13: 978-0123704733.

[22] John Askill, *Physics of Musical Sounds* (Van Nostrand Reinhold Inc., 1st ed., 1979). ISBN-13: 978-0442203818.

[23] Arthur H. Benade, *Fundamentals of Musical Acoustics*, 2nd Revised Edition (Dover Pub., 1990). ISBN-13: 978-0486264844.

[24] Diana Deutsch, Ed., *The Psychology of Music*, 3rd ed. (Academic Press, NY, 2013), http://www.sciencedirect.com/science/book/ 9780123814609

[25] Donald E. Hall, *Musical Acoustics: An Introduction*, 3rd ed. (Brooks/Cole, 2001). ISBN-13: 978-0534377281.

[26] William M. Hartmann, *Principles of Musical Acoustics* (Springer, 2013). ISBN 978-1-4614-6785-4.

[27] Donald Gray Miller, *Resonance in Singing: Voice Building Through Acoustic Feedback* (Inside View Press, 2008). ISBN-13: 978-0975530757.

[28] John S. Rigden, *Physics and the Sound of Music*, 2nd ed. (J. Wiley, 1985); ISBN-13: 978-0471874126. William J. Strong and George R. Plitnik, in *"Music Speech High-Fidelity"* (Sound print, 1983). ISBN-13: 978-0961193805.

[29] Thomas D. Rossing, Ed. *The Science of String Instruments* (Springer, 2010); DOI: 10.1007/978-1-4419-7110-4; Antoine Chaigne and Jean Kergomard, *Acoustics of Musical Instruments* (Springer, 2016). ISBN 978-1-4939-3677-9.

[30] William R. Savage, *Problems for Musical Acoustics* (Oxford University Press, NY, 1977). ISBN-13: 978-0195022513.

[31] Johan Sundberg, *The Science of the Singing Voice* (Northern Illinois University Press, 1989). ISBN-13: 978-0875805429.

[32] S. Daniels, T. Kodama, D. J. Price, Ultrasound in Medicine & Biology, **21**, 105 (1995).

[33] Federal Highway Administration Transportation Noise Model (https://www.fhwa.dot.gov/environment/noise/traffic_noise_model/).

[34] L. Beranek, Noise and Vibration Control (Inst. of Noise Control; Revised edition, 1988). ISBN-13: 978-0962207204; see also James P. Cowan, *Handbook of Environmental Acoustics* (John Wiley & Sons, 1993). ISBN: 978-0-471-28584-7.

[35] M. P. Norton, D. G. Karczub, *Fundamentals of Noise and Vibration Analysis for Engineers* (Cambridge University Press, 2003). ISBN-13: 978-0521499132

[36] H. Panofsky and J. Dutton, *Atmospheric Turbulence: Models and Methods for Engineering Applications* (Wiley Interscience, NY, 1984).

[37] NIDCD (Nat. Inst. on Deafness and Other Comm. Disorder) https://www.nidcd.nih.gov/

[38] NIOSH, "Criteria for a Recommended Standard–Occupational Exposure to Noise" (National Institute of Occupational Safety and Health, 1972).

[39] OSHA noise regulation: https://www.osha.gov/dts/osta/otm/new_noise/

[40] WHO: Occupational exposure to noise, B. Goelzer, C. H. Hansen and G. A. Sehrndt, Eds., http://www.who.int/occupational_health/publications/occupnoise/en/; WHO: Environmental Noise Guidelines for the European Region (2018). ISBN 9789289053563.

Index

Printed in the United States
by Baker & Taylor Publisher Services